About Island Press

Since 1984, the nonprofit organization Island Press has been stimulating, shaping, and communicating ideas that are essential for solving environmental problems worldwide. With more than 1,000 titles in print and some 30 new releases each year, we are the nation's leading publisher on environmental issues. We identify innovative thinkers and emerging trends in the environmental field. We work with world-renowned experts and authors to develop cross-disciplinary solutions to environmental challenges.

Island Press designs and executes educational campaigns, in conjunction with our authors, to communicate their critical messages in print, in person, and online using the latest technologies, innovative programs, and the media. Our goal is to reach targeted audiences—scientists, policy makers, environmental advocates, urban planners, the media, and concerned citizens—with information that can be used to create the framework for long-term ecological health and human well-being.

Island Press gratefully acknowledges major support from The Bobolink Foundation, Caldera Foundation, The Curtis and Edith Munson Foundation, The Forrest C. and Frances H. Lattner Foundation, The JPB Foundation, The Kresge Foundation, The Summit Charitable Foundation, Inc., and many other generous organizations and individuals.

The opinions expressed in this book are those of the author(s) and do not necessarily reflect the views of our supporters.

Science with Impact

Science with Impact

HOW TO ENGAGE PEOPLE, CHANGE PRACTICE, AND INFLUENCE POLICY

Anne Helen Toomey

Copyright © 2024 Anne Helen Toomey

All rights reserved under International and Pan-American Copyright Conventions. No part of this book may be reproduced in any form or by any means without permission in writing from the publisher: Island Press, 2000 M Street, NW, Suite 480-B, Washington, DC 20036-3319.

Library of Congress Control Number: 2024934827

All Island Press books are printed on environmentally responsible materials.

Manufactured in the United States of America
10 9 8 7 6 5 4 3 2 1

Keywords: citizen science, civic science, communication and science, evidence-based policy, evidence-informed policy, inclusive science communication, open science, parachute research, participatory science, peer review, research ethics, research methods, research practice gap, research with impact, risk communication, science communication, science education, science skeptics, science-society interface, scientific impact

For David, who loved birds, trees, and, most of all, people

Contents

Introduction: Science—The Next Generation   1

Part One: Searching for Impact
Chapter 1 Will You Please Just Listen to Me?   13
Chapter 2 Will I Please Just Listen to You?   33
Chapter 3 From Impact to Encounter   51

Part Two: The Spaces of Scientific Impact
Chapter 4 Asking a Good Question   73
Chapter 5 The Privilege of Choice: Methods, Permissions, and Location   95
Chapter 6 The Power of Participation: Data Collection and Analysis   117
Chapter 7 Rethinking the "Peer" in Peer Review   139

Part Three: The End Is Just the Beginning
Chapter 8 The Scientist Next Door: Conversations, Communities, and Connections   161

Chapter 9 The Skeptic in the Mirror: The Essential Role of
Uncertainty in Science   179
Chapter 10 In the Belly of the Beast: Scientists, Policy-Making,
and Advocacy   201

Conclusion and Acknowledgments: From Boldly Going to
Steadily Engaging   221

Notes   227
Index   277
About the Author 285

INTRODUCTION

Science—The Next Generation

I DISCOVERED I WAS A STAR TREK NERD relatively late in life, in my midthirties. I amused my family members with regular references to Vulcans and Klingons, androids and warp drives. Before watching *Star Trek: The Next Generation*, I had assumed that a Trekkie (or Trekker, the more distinguished term) was someone more obsessed with space than with life on Earth, but as I became increasingly immersed in the world of the USS *Starship Enterprise*, the appeal that the show held for me was not in its discovery of new worlds and alien life-forms, but in the promise of what human society could someday become—equitable, advanced, and plentiful. Humans in *The Next Generation* were no longer plagued by twenty-first century problems such as poverty, discrimination, and war, in large part because the promises of science and technology ensured that the basic needs of all humanity had been met. In this bold, new world, all individuals were thus free to pursue their passions and curiosity.

As I got over my initial embarrassment of being a Trekker, I came to learn that most of my scientist friends were also *Next Gen* fans. How could one not be? The series—more than any other—validates what science is about.[1] The dream of the rational, peaceful, and supremely intelligent human and alien creature (if they happen to be Vulcan, anyhow) is both reassuring and compelling. Multidisciplinary research teams explore the universe in search of new ideas, friendships, and, ultimately, meaning. Contact and the ensuing collaboration with other cultures are highly esteemed; meals are eaten in a mess hall with colleagues and friends; and every member of the crew is valued not just for their individual genius, but for how they contribute to the greater good of the mission. Perhaps most importantly, science's role as a beneficiary goes unquestioned, not only for humanity, but across the reaches of the galaxy.

Contrast this utopian vision with the relationship that science holds with society today—one in which the scientific consensus on key issues, including climate change, vaccines, and, in some corners, even of the roundness of Earth—is frequently called into question.[2] Sometimes it seems as if the more evidence that mounts, the more pushback it faces. And such perceptions are not shadows in the dark. They result in major consequences for society, such as the battles over the teaching of evolution in some parts of North America or the documented decrease in routine childhood immunization across the world.[3]

Those of us who work in science increasingly believe that it is our responsibility to do what we can to explain its value. We write blog posts, we give radio interviews, we get into heated debates on social media, and we spend countless hours preparing lectures for undergraduates who, we hope, will lead us one step closer to a *Next Gen* future. Yet it often seems that no matter what we do, it isn't really helping. Even as we produce more and more reliable data, we are beginning to understand that facts aren't quite enough, and we're at a loss about what to do next.

I understand this feeling because, as someone who has worked in the environmental field as a researcher and practitioner for more than twenty years, it has long played in my own mind. Scientific research has been crucial to our movement—from measuring air and water pollution to studying the benefits of clean environments for physical and mental well-being. I got into research to better understand how people perceive, connect to, and ultimately care for natural spaces, whether those be richly biodiverse landscapes of the Amazon or polluted waterbodies in densely populated cities. But I never wanted my research to stay within the confines of academia. I wanted it to *matter* for society—to have impact in terms of the way people thought or even acted. And I know I'm not alone.

Many of us in the scientific community are here because we care about producing interesting research and we wish to "make an impact" with our work. We want to influence the way society thinks about a certain issue and, in turn, the decisions of policy makers. But often we don't feel like we know how to achieve those goals. Scientists endure criticism for not doing more to educate or engage with the public. Almost within the same breath there is acknowledgment that scientists lack the training and skills to communicate effectively. Even James Hansen, the NASA scientist and climate activist known for his groundbreaking testimony to the US Congress about climate change in late 1980s, said, "Most scientists are not good communicators. We're not trained to do that. Science itself is hard enough without trying to communicate with the public."[4]

He has a point. As a newly tenured professor, I have a deep understanding of the pressures in academia that make it almost impossible to take the time needed to think about how my research could have a greater impact on society, let alone to engage in actions to do so. I know the stress that comes with trying to juggle an ever-increasing load of teaching, research, and administrative tasks and the frequent fantasy of giving it all up to start a sheep farm in Vermont. I can remember once, during a particularly frenzied hour as a new assistant professor, eating my lunch at my desk so fast that a piece of bread flew out of my sandwich and got stuck in my eye.

In part because of the sheer amount of work that needs to be done, academic research is increasingly lonely and stressful. As a result, we are entering and leaving academia in droves—pulled in by its promise, pushed out by its pressures. Even among those who have "made it," some are looking back at their careers and wondering if the impact of their science is what they intended at the start. A good friend of mine, a tenured full professor of ecology at a research institution, told me that he's taken to emphasizing to students the importance of fields like business and engineering in addressing environmental issues like climate change, as opposed to specializing in science-focused disciplines such as ecology. "We don't need any more papers about the predator-prey relationships of dragonflies and tadpoles," he told me, provocatively, having published quite a few himself.

But I have pushed back against my friend, not because I think he is wrong about adding to the stack of papers on dragonflies and tadpoles but because I think there is a future for those who wish to make a difference with their research. However, something different is needed than the model we currently have, which is largely based on the premise that the fact-based recommendations made by scientists will somehow trickle down to the masses and influence the way our societies function.

This is where I'll return to *Next Gen*. At first glance, this utopia seems to be a natural product of three centuries of technological progress and scientific discovery that has made life easy and plentiful. The famous *Star Trek* "replicators" can instantly produce anything—food, water, clothing, the latest Barbie doll—anyone could desire for free, thus obliterating the need to earn money. Faster-than-light-speed travel (warp drives) enable humanity to connect with other civilizations and thus broker world (and, ultimately, intergalactic) peace. And medicine is constantly advancing in leaps and bounds, with "hyposprays" instantly administering vaccines to newly discovered diseases, basically eliminating illness.

Upon deeper analysis (cue to rewatch those favorite episodes), however, what makes the *Next Gen* utopia so compelling is not really the advanced gadgets and gizmos. Yes, warp drive, food replicators, and medical tricorders are cool, but they don't automatically make for a better world. Rather, what makes all the difference are the *choices* made around science and technology.[5] Intergalactic travel could have meant colonization of other planets and an expansion of war and exploitation, as in many other sci-fi franchises, but the *Enterprise* is not a military vessel but a scientific one, dedicated to encountering (not conquering or extracting) other forms of life and knowledges. Food replicators are not patented, profit-driven ventures, but are open-source technology, shared freely among more than one hundred planets.

In the *Star Trek* universe, choices are also made about whose knowledge counts, which reflect certain values. Yes, logic is prized (after all, who doesn't love our pointy-eared friends?). But valid knowledge and valuable perspectives come in many shapes and sizes and from many lifeforms and cultures, including those not human. Diversity of all kinds is seen as extremely valuable, and different abilities are not just accommodated, but celebrated.[6] The autonomy of alien worlds is protected by a commitment of noninterference (the "Prime Directive"), so societies are left to develop in the ways they see fit. The most interesting discoveries (and episodes) arise when a given character has been challenged to see things differently and thus must update their worldviews and actions accordingly. Ethics and philosophy are constant companions of science and technology, seen as friends rather than foes.[7]

So, too, is this book about choices. Specifically, it is about the choices that scientists and their supporters can make to effectively engage with communities, influence policy, and get more people genuinely excited about science. Years ago, I began a journey to tackle, head on, the question of what makes research impactful for society. Throughout this process, I interviewed many people from different walks of life—environmental managers, policy makers, Indigenous land stewards, farmers, educators, health workers, science communicators, agricultural extension agents, and lobbyists, among others—about their perceptions of research and the role of evidence in individual and collective processes of change.

At the heart of my inquiries is the word—or perhaps concept of—*impact*. Dictionary definitions of the word often include phrasing such as "an impinging or striking especially of one body against another" and "the force of impression of one thing on another." Synonyms include bump, collision, concussion, crash, impingement, jar, jolt, jounce, kick, shock, slam, smash, strike, wallop, affect, impress, influence, move, reach, and sway. If you search images of "impact" online, one common image you will find is of a drop of water splashing into a larger body of water, ripples spreading outward. Another is a finger about to topple a line of dominos, where they will fall in sequence, one after the other.

These definitions and images convey a few popular notions about what we think impact means. First, that for impact to occur, two or more separate entities are required, one that impacts the other in a predetermined direction—an impactor and an impactee. Further, there is the impression that an impact is something controllable. We throw the stone, we make the ripple—the larger the stone and force, the greater the ripple and the longer and farther the waves will continue to go. Similarly, impact is often seen as predictable. If we push one domino and the others are lined up and ready to go, they too will fall in a predetermined row.

Now more than ten years into my journey, one thing is clear: scientific research can—and frequently does—have great impact for society, but not in the way many of us have traditionally believed. As you will learn in the coming chapters, impact is not straightforward, nor does it travel in one, predictable direction. Rather, it looks more like a complex network, with arrows and lines spreading across space and time and where scientific "facts" are just one piece of a very large system that, above all, is social in nature.[8]

In this book, I will share what I have learned about the impact of science beyond academia to influence how policy makers make decisions and how society operates. I will share some of my own choices, both good and bad, to demonstrate the types of impact (not always positive) that can arise at different stages in the scientific process. I will also talk about how my choices were never just my choices, but rather emerged from a complex mix of individual preferences, societal pressures, and institutional resources (or the lack thereof). As such, this book is written not only for the individual researcher, but for a wider professional audience committed to the societal use of research, including those involved in grant funding, science communication, and research policy. The choices scientists make about how to "have impact" are thus not theirs alone but belong to the entire scientific community.

I made a few choices while writing this book. First, I opted to publish with Island Press, a small, nonprofit publisher that specializes in environmental and science books, rather than a standard academic press. I did so because I wanted to reach a wider, professional audience, not just an academic one. I also wanted some freedom to have fun with the writing without going down too many scholarly rabbit holes.[9] For those who wish to dive further into the scholarship, I have provided an extensive endnotes section with resources.

Additionally, I have tried to make this book as much of a guide as possible. To this end, part one partially follows my own journey in learning more about the different ways science can have impact. Chapter 1 describes the way I used to think science communication had to work, which I like to call the "Will You Please Just Listen to Me" approach. Chapter 2 provides an alternative model of how to spread science-based ideas and behaviors based on the story of one of the most powerful public health achievements in history. And chapter 3 describes how my understanding of impact changed from focusing on the products of science to focusing on a process that is based in opportunities for encounter and connection.

I have also structured some of the chapters the way my friend and colleague Monica Palta recommended. She was trained in ecology and biology and suggested that the book should be structured the way scientists think—like the different stages of the scientific process. That made a lot of sense to me, and part two is organized in this fashion. For example, chapter 4 focuses on the research stage when we ask questions; chapter 5 is about choices we make when setting up a research project, such as determining location and methods; chapter 6 is centered on the data collection stage; and chapter 7 addresses peer review and the dissemination of research results. In this sense, the book takes a closer look at the multitude of choices that are made throughout the process of scientific inquiry, even before any samples are collected or analyzed.

When it comes to impact, what many people often think about is disseminating the results of science to policy makers and various other "publics." That's the focus of part three. Chapter 8 explores conversations and stories about science, chapter 9 dives into the question of how to communicate uncertainty, and chapter 10 is about engaging in policy and advocacy. The chapters in parts two and three have accompanying flowcharts on islandpress.org/impactful to guide readers to the specific choices they might need to consider in various stages of their journey.[10]

Although this book provides guidance on including and impacting a wider array of participants in science, there are important aspects of diversity and inclusion in science that it does not directly address, such as how to increase participation of people who have been historically excluded from the science and technology workforce, particularly women, minorities, and persons with disabilities. For more guidance on how educators, mentors, researchers, and academic administrators can create equitable opportunities for these groups, two excellent books are *Making Black Scientists* by Marybeth Gasman and Thai-Huy Nguyen and *Women in Science Now* by Lisa M. P. Muñoz.[11] Similarly, this book does not directly explore the various types of impacts that science can have in terms of industrial, technological, and military uses and related ethical debates. A superb book that does directly tackle the broader issue of values in science for industry and other considerations is *A Tapestry of Values* by Kevin Elliott.[12]

Some choices I made around the guidance provided in this book were not quite my own but, rather, stem from my positionality and location in the world. I write from the village of Sleepy Hollow, about an hour's drive north of New York City, where the institution that employs me, Pace University, is located. Although I have spent approximately half my adult life outside the United States (in Mexico, Nicaragua, Costa Rica, England, and Bolivia), my perspective is very much influenced by the political, ethical, and social debates happening in my neck of the woods.[13] Thus, while I have attempted to provide examples from other parts of the world in this book, its message comes to you from a specific lens—one that is North American, white, female, and politically liberal. As any responsible researcher, I've shared drafts of this book with many different people, all of whom have their own set of lenses through which they view the world, and it is my hope that this practice, more than any other, will make the material relatable, interesting, and applicable to different contexts.[14]

Above all, I chose to write a book that would reconcile the idealistic Trekker within and the logical pragmatist who dives into the peer-reviewed literature when she has a question about how the world works. Fundamentally, this book is written to challenge the scientific community to consider whether our work is having the desired impacts for society and, if it is not, how we can change to make it so. Although calls for change can seem daunting in the face of current barriers, structures, and incentives in the academic system, in the pages ahead I hope to show that such changes are not only possible, they're already happening. Or to put it in the defiant words of Jean-Luc Picard, captain of the USS *Enterprise*, "Things are impossible until they're not!"[15]

PART ONE

Searching for Impact

"Once you do away with the idea of people as fixed, static entities, then you see people can change, and there is hope."
—bell hooks

"If you think you're too small to have an impact, try going to bed with a mosquito."
—Anita Roddick

CHAPTER 1

Will You Please Just Listen to Me?

IF *STAR TREK: THE NEXT GENERATION* provides an idyllic vision of what humanity could become if science takes its rightful place in society, an increasing number of apocalyptic films seek to portray what will happen if it does not. One film in this category, 2021's *Don't Look Up*, features two astronomers—PhD candidate Kate Dibiasky and Dr. Randall Mindy (played by Jennifer Lawrence and Leonardo DiCaprio, respectively)—who discover that a comet will collide with Earth within the year, causing the extermination of the human population.[1] However, when they present their findings to White House officials, they are met with derision and apathy. The film quickly devolves into an Earth-Is-Doomed adventure complete with internet memes, sex scandals, and an Elon Muskesque scheme to harvest rare elements from the comet instead of destroying it. It's quite a ride, and the message is clear: look what happens if we don't listen to scientists.

Although *Don't Look Up* is clearly a satire, designed to exaggerate the apathy of government officials and ignorance of the public to very real threats, it was explicitly written as an allegory for how governments, media, and the wider public have reacted (or not reacted) to decades of warnings from scientists about climate change. One article published in *Wired* magazine, aptly titled "*Don't Look Up* Nails the Frustration of Being a Scientist," quoted climate scientists who said they could relate to the experiences of the main characters, who were simultaneously desperate to get the word out and baffled at why their message was not getting across.[2] In another article, the film's scientific advisor, NASA scientist Dr. Amy Mainzer, was quoted as saying, "I told (the actors), you really have to speak for the science community, when we feel like we're being ignored. . . . There are some lines in the movie that are just, 'we're trying to tell you, you're not listening and we really want you to listen, because we know we can make things better if everybody does.'"[3]

The idea that scientists need to be listened to is at the heart of what many of us think of when we talk about science communication and scientific impact. Rigorous scientific research is carried out, again and again, until the scientific community begins to develop a consensus around a topic. The science is so strong that it is practically indisputable (in the case of *Don't Look Up*'s comet, a 99.78 percent chance that it will destroy Earth). Furthermore, the science has clear policy implications—something must be done to address the issue. Scientists and their allies thus must do what they can to get the right information to the right people to make this change.

I like to call this model the "Will You Please Just Listen to Me" approach to science communication. It assumes that the problem is one of lack of knowledge, understanding, and motivation to act and thus what is needed is to get the facts out to the masses to convince people to do what the science says needs to be done.[4] Indeed, in the film, the scientists tried their darndest to do just that. They repeatedly talk to the press, they post livestream videos on social media, they travel to Washington, DC, to speak with policy makers, and they even organize a benefit concert with rapper Kid Cudi and singer-songwriter Ariana Grande, who emerges onstage dressed like a comet to perform the original song "Just Look Up" for tens of thousands of lighter-holding fans.

Alas, their efforts are not exactly met with resounding success. The scientists, particularly the young female PhD student, are ignored and mocked, first by White House staff, then by the media, and even within their own social circles. In one humorous scene, the scientists wait for hours outside the Oval Office for a reception with the president of the United States and are charged for snacks by an Air Force general. Later in the film, Dibiasky, the student who discovered the comet, goes home to find that her parents have locked her out of the house. "No politics!" her dad says as they look at each other through the screen door. "We're for the jobs the comet will provide," says her mother.[5]

Thus, even as the scientists spend their time and money trying to save the world, nothing is enough. Spoiler alert: Humanity will not save itself from the comet, though at least the scientists and their allies get to have one good final meal before they die. So, although *Don't Look Up* does a great job at demonstrating that the "Please Just Listen to Me" approach isn't sufficient to convince society to take action, it doesn't quite explain why that is the case and what scientists might do differently.

So why aren't the facts enough to incite change or influence policy? How can it be that large groups of individuals see some scientific findings to be in conflict with their worldviews, thus forestalling societal change and political action? To begin to answer these questions, let's travel to Tangier Island in Virginia, a tiny island that is quickly going under water due to sea level rise and whose residents are skeptical that climate change is the culprit.

Notes from a Disappearing Island

Tangier Island is a small island in the Chesapeake Bay off the eastern coast of the United States, part of the state of Virginia. For centuries, the island was used by the Pocomoke peoples for oystering, fishing, and possibly as a seasonal settlement.[6] In the eighteenth century, the island was homesteaded by White farmers, who established a community that eventually grew to approximately thirteen hundred inhabitants by 1930.[7]

Along with many islands in the Chesapeake Bay, Tangier Island is slowly being swallowed by the sea. Since 1850, two-thirds of the island's landmass has disappeared, a trend that has been increasing due to erosion driven by sea level rise and increasing storms.[8] As the land has disappeared, so have many of the inhabitants; by 2021, the population of Tangier had dwindled to approximately four hundred residents.[9] Scientists say that climate change is the main culprit and predict that the island will be uninhabitable within the next few decades.[10] If that happens, the residents of Tangier will be among the first climate refugees in the United States, a label more commonly associated with island communities in the South Pacific, who have been at the front lines of the fight for climate action at an international level.

However, unlike communities from the Pacific islands of Tuvalu, Nauru, and Kiribati, the residents of Tangier have not been spokespeople for climate action. To the contrary. Many of the residents of Tangier contest the idea of human-caused climate change and sea level rise and think that it is, if not quite a hoax, a distraction from the real issue of God-made erosion. Indeed, if you travel to Tangier, you can stop by a gift shop and purchase a T-shirt that reads, "I refuse to be a climate change refugee."[11]

Scientists often understand that scientific knowledge on a particular issue is difficult to get across to nonexperts. It's not always easy to explain the mechanisms of how things work, especially when those mechanisms are abstract, invisible to the human eye, or beyond one's personal experience.

But the impacts of climate change on Tangier Island are none of these things. Residents have regular floods in their yards and in their homes. The level of the Chesapeake Bay has risen by approximately three feet in the last five hundred years, the number of intense storms has increased, and whole sections of the island have been lost along with the homes of many residents.[12] Although many communities across the world are facing impacts from climate change, Tangier is a place where it can be experienced in a visceral way.

But that is not how many Tangier residents see it. In 2017, the island became the center of a media storm when its mayor, James "Ooker" Eskridge, was filmed making a plea to then US president Donald Trump for help in shoring up their island. "They talk about a wall; we'll take a wall," said Eskridge, adding that he was a fan of Trump.[13] In an atmosphere in which much of the country was still reeling from Trump's election to office and his quick withdrawal of the United States from the Paris Climate Agreement, the subsequent public shaming was quick, witty, and mean. In the opening monologue of *The Late Show with Stephen Colbert* in June 2017, Colbert made fun of the mayor's faith in Trump to cut through red tape to protect the island, saying, "Trump is going to get them that wall and make the ocean pay for it!"[14] And in another news satire program, *Full Frontal with Samantha Bee*, a reporter visited Tangier Island to interview its residents, poking fun at the social, cultural, and religious differences between the islanders and her (presumably) more urban and liberal audience.[15]

What was lost on the media, who often reported this story as a battle of conservative, religious viewpoints over liberal, science-based perspectives, was that the islanders of Tangier simply, and quite literally, saw things differently. When residents said it was erosion, not climate change, taking their island, they were not just being obstinate. They really meant it.

What Our Brains Do in the Shadows

Think about these questions:

> Do you believe we are already experiencing the impacts of climate change?
> Do you believe human activity is the primary cause of modern-day climate change?
> How did you determine your beliefs?

If you have a few minutes, go online and research these climate change topics. Perhaps you will find data and information to support your beliefs, and perhaps you will also find data and information that challenges them. Which did you click on? Which did you spend the most time reading?

Now think about these related questions:

> Why exactly does the presence of carbon dioxide in the atmosphere create a greenhouse effect?
> If pressed, could you explain why to someone else?

If you are not an atmospheric scientist and find these questions somewhat challenging, take a moment for some online research. Possibly you will find some websites that refer to the molecular structure of various molecules and how the structure of carbon dioxide differs from that of nitrogen gases. Okay, that might make sense. Read on, and you may learn about how the bonds between atoms in certain molecules can vibrate in different ways, which influences whether the energy of a photon corresponds to the frequency of a molecule and then whether that energy (or heat) is absorbed. If the infrared photon is absorbed, the energy will cause the molecule to vibrate—until it in turn reemits the photon to another molecule, increasing the speed of that other molecule's motion.

Have you had enough yet?

If you are like most people, with limited time and your own work and life concerns, after a while you will probably conclude that you don't have time to understand the scientific details that underpin the greenhouse effect and settle on one or two articles from sources you trust. That's completely natural and normal. The world is extremely complex, so the way we make sense of it is to rely on what our intuition, the world around us, and other people tell us to be true.

In research, we do that all the time. We don't know every aspect of what we study, yet that does not stop us from collaborating on projects and papers with experts from other disciplines whose expertise fills in our own gaps. It's not just because science is overly specialized; it's because there is an extraordinary amount of information out there and our brains can only contain so much information. This is true for all of humanity and can be explained through our evolutionary past.

Humans are social creatures. We evolved from hunter-gatherer units in Africa, and in many ways we were not physically well adapted to our environment. We could not run as fast as animals that preyed on us or those we preyed on. Our skin was not covered in protective spines or scales, and we suffered greatly from cold temperatures. We needed to drink and eat every day, and many foods available to other animals were poisonous to us. Our young were born so developmentally immature that we had to watch them carefully for years lest they be snatched up by a lion or bitten by a snake.

Given that early humans were comparatively weak, vulnerable, and slow, how did we not only survive, but thrive, in the face of such handicaps? The answer is deceptively simple: we learned to cooperate.[16] We formed social groups that shared the labors of hunting, building shelter, preparing food, and rearing children. It was a lot of work, so we became specialized in certain tasks and relied on others in our group to become experts in others. Some group members hunted while others gathered medicinal plants and prepared meals, which allowed those who were hunting to dedicate time and effort toward improving their skills in that area and, similarly, for those preparing the food to make it taste better.

As we have evolved as a species and our society and technology have become more and more complex, our knowledge has become increasingly specialized.[17] We may know how to go to the supermarket and buy the correct ingredients to make chili for our friends, but very few of us know how to raise and butcher a cow for ground beef. We live in homes full of piping and wires that we do not fully understand, so when the sink backs up, we typically call a plumber, and when a fuse blows, we call an electrician. In our daily lives, we rely on the expertise of others so regularly that we almost forget how reliant on others we are. As cognitive scientists Steven Sloman and Philip Fernbach argue in *The Knowledge Illusion: Why We Never Think Alone,* we live in the illusion that we are independent thinkers and self-sufficient, an illusion that holds as long as the internet doesn't inexplicably cut in the middle of watching our favorite football team.

Because we cannot know everything, when we encounter new knowledge, we do our best to make sense of it. But "making sense" of that knowledge does not mean some Spock-like approach to logical information processing. Rather than carefully deliberating over every new piece of information, what typically happens is something more akin to a gut reaction. Our brains automatically make assumptions about the validity and importance of the information, relying more on our prior experiences than logic to inform us.[18] Most of this process is unconscious and requires very little effort or energy, which is by design. Our brains evolved to prioritize survival. If we heard noises in the underbrush that sounded like a predator approaching, we didn't wait around to confirm our assumptions—we ran for it.

Behavioral scientists often refer to our brains as "cognitive misers," cheapskates when it comes to having to spend time or conscious thought on new ideas.[19] Being cognitive misers means that we often rely on almost anything besides brain effort to make sense of the world, and thus we typically use mental shortcuts (often referred to as heuristics) to make rapid assessments about ideas about which we know very little.[20] Being a cognitive miser is mostly a good thing as it means that we can go about our lives efficiently, not spending hours deliberating the utility of everyday actions. Life without such heuristics would be exhausting—we would regularly spend hours pondering matters as trivial as what to have for breakfast or the best route to get to work.

The use of such mental shortcuts can have costs, however. Specifically, they bias our decision-making processes, causing us to get stuff wrong. Our cognitive stinginess means that we tend to rely heavily on previous experiences and stories that quickly come to mind for decision-making rather than taking the time needed for thoughtful deliberation. We're also prone to see patterns that aren't there and to accept or refute new information without evidence. And "we" isn't just other people—it is all of us, scientists included. For example, historians have argued that some of the most prominent examples of science losing its way have come about when the leading researchers at the time were biased toward one theory or approach, discounting others.[21]

In addition to our cognitive limitations, another major influence in terms of how we interpret new information is related to our social networks and identities. As mentioned earlier, humans are social animals, and if we don't know what to think about something, we often turn to those we trust to tell us what's what.[22] We are attuned to what is considered normal or acceptable in our social environments, and we instinctively know that acting in violation of such norms will have negative consequences. Most such social obedience feels so natural that we don't even realize we're doing it most of the time. Typically, it's only when we encounter someone who doesn't share a given social norm that we realize our behavior is learned rather than natural. For example, an Indian friend shared that she was taught to touch elders' feet as a sign of respect, which was surprising to me, as doing so among my own family would be disconcerting to all parties.

Prior experiences and social factors are hugely influential in how we see the world, sometimes quite literally so. One striking example is "the dress" phenomena, in which a picture of a blue-and-black dress was posted on social media. Bizarrely, not everyone's brains processed the colors of the dress in the same way. Some people, including myself, saw the colors of the dress as being white and gold. Neuroscientists who later studied the phenomena discovered that what was happening was that the picture was so overexposed it essentially forced our brains to make assumptions about the lighting conditions in which the photo was taken.[23] Because our perceptions of color are informed by our perceptions of lighting, those of us whose brains assumed the picture was taken under natural lighting (possibly outside) saw it as white and gold, and those whose brains assumed it was taken under artificial lighting (indoors) saw it as blue and black.

Researchers surveyed thirteen thousand people to try to understand why some brains assumed one lighting condition and not another and found that the main difference between the blue-black people and the white-gold people was how much time in their life they had spent working inside as compared to outside. Those who were more accustomed to outdoor lighting conditions were more likely to see the dress as having been illuminated by natural light, which led their brains to subtract blue light, leading the image to look more yellow. In other words, our brains quite literally saw the dress as being a different color based on our prior life experiences. This process is out of our control. For example, even though I now know that the actual color of the dress is blue and black, I still see it as white and gold. I've tried to tell my brain that it is wrong, but I suppose my years of working (and even living) outdoors will simply not allow me to see it differently.

This example is rare in some ways, but in another sense, it is extremely common. In his book *How Minds Change*, David McRaney used the example of "the dress" to explain that just as our brains can trick us into seeing different colors or shapes based on a process of filling in missing information, so do we do this in other areas of our life.[24] When faced with a novel situation, our brains will automatically and unconsciously make assumptions based on what we have most frequently encountered in the past and what "feels right" based on our social environments. For example, researchers at Yale's Cultural Cognition Project conducted a study in which they showed two separate groups of participants a video of people protesting and asked them whether they believed the protesters were acting lawfully or not. One group of participants was told that the protesters were protesting abortion outside of an abortion clinic, and another group was told that the protesters were protesting against the military's then-existing position against openly LGBTQ+ soldiers outside a military recruitment center. The researchers found that participants' political and cultural identities largely dictated how they interpreted the actions of the protesters.[25] Participants who the researchers categorized as having more "egalitarian and individualistic" worldviews (associated with more liberal or progressive perspectives) saw the abortion protesters, but not the military protesters, to be unlawful, whereas "hierarchical communitarians" (associated with more conservative or traditional viewpoints) perceived the opposite. But they were all looking at the same video. This tendency to interpret the same information differently, McRaney argued, is why we disagree so much and why such disagreements often seem to happen across different social identities (such as political ideologies), writing: "Unaware of the processing that leads to such disagreement, it will feel like a battle over reality itself, over the truth of our own eyes. Disagreements like these often turn into disagreements between groups because people with broadly similar experiences and motivations tend to disambiguate in broadly similar ways, and whether they find one another online or in person, the fact that trusted peers see things their way can feel like all the proof they need: they are right and the other side is wrong factually, morally, or otherwise."[26]

Because of this reaction, information that challenges our preexisting worldviews or our social identities can feel threatening and even offensive. The more complex and stressful the information, the more likely it is to be processed in parts of the brain such as the insula, ventral striatum, or amygdala (associated with emotions such as fear or pleasure) rather than in the prefrontal cortex (associated with rational thinking and deliberation).[27] Research has found that although winning an argument triggers "feel good" hormones such as dopamine and adrenaline, when our beliefs are threatened, we release cortisol, a hormone associated with stress and the fight-or-flight response.[28]

In a similar way, climate change was not a neutral concept to Tangier residents. For decades, they had been trying to raise awareness about the increasing rate of land loss to the sea. In fact, the islanders had been doing measurements of their own for more than a half century, calculating the loss of land from the ocean long before scientists arrived to talk about climate change.[29] That their island was disappearing, and disappearing fast, was something anyone who lived on Tangier Island knew. Erosion was part of the natural cycle of life, which fit into their Biblical worldviews that God controls the doings of Earth. Furthermore, erosion was the so-called enemy they knew, something that their parents and grandparents had dealt with and something that could be addressed in the future.

In contrast, climate change was a concept from the outside, promoted by the "come-heres" ("outsider" in the local dialect of the island) with whom the residents had little in common. It was human caused and often used by liberal politicians (the "other" political party) to promote solutions such as carbon taxes and electric cars, but who offered little in terms of protecting coastlines. Whereas erosion could be addressed by shoring up the perimeter of their island—perhaps by a seawall—climate change meant the end of their island.

Furthermore, scientists had a poor track record as far as Tangier residents were concerned. The residents had a long and frustrating history trying to work with scientists and government officials to bring actionable solutions to address the loss of land. The first government study commissioned to look into the problem was conducted in the 1970s, and solutions to protect the island have typically been half measures that proceed at a glacial pace. In 2017, when the press visited the island in droves to poke fun at the irony of a climate-denying island of soon-to-be climate refugees, islanders were waiting for construction of a promised rock jetty to begin. The jetty had first been proposed in the mid-1990s but had been mired for years in studies and red tape. In his book *Chesapeake Requiem: A Year with the Waterman of Vanishing Tangier Island*, author Earl Swift captured the frustration of the townspeople, epitomized in a conversation he had with Mayor Eskridge, who said, "They do studies, then they study the studies. I know that's their procedure, but it gets frustrating. We're at the point now that it's like me coming across a family in a boat that's sinking, and I say, 'I'm going to rescue you, but I have to study it first.'"[30]

How a Hockey Stick Became a Boomerang

In 2017, with all the press attention on Tangier Island, Eskridge was invited to take part in a televised town hall on climate change, which featured former vice president Al Gore.[31] Gore is most well known for raising awareness about climate change through the release of his award-winning documentary film, *An Inconvenient Truth*.

I clearly remember when *An Inconvenient Truth* came out. It was 2006, and I was in a graduate program focused on environmental issues. Climate change was a regular point of discussion in my classes, among my peers, and with my professors. Prior to watching the film, I had already accepted the links between an increase in greenhouse gases, global temperatures, and more intense and more frequent hurricanes in wetter parts of the world and unprecedented wildfires in drier parts. My left-leaning worldview was also consistent with the policy recommendations that would inevitably stem from climate science—reductions in CO_2 and other greenhouse gas emissions from polluting corporations and a focus on community and social responsibility for addressing the issue.

So as a budding environmental researcher, Gore's film really resonated with me. At the time, I remember thinking, "Wow, now everyone will know about this problem." To my twenty-something mind, Gore was straightforwardly presenting the facts, even using graphs to prove his points. Who could dispute that?

Many people did. Conservative-leaning Americans saw Gore, a Democrat who lost the 2000 US presidential election and thus failed to advance his policies through government, make a film depicting a global environmental problem that could only be solved with sweeping policy change. More taxes. More environmental regulation. More public transportation. Hmm. Such proposals were like those that had been promoted for decades by liberals, long before discussions of climate change took the national stage. Right-wing media called the phenomenon a "watermelon"—green on the outside, red on the inside. They mocked climate change action as an "Al Gore deal" and flouted the hypocrisy of his flying around the world to promote the issue, charging a hundred thousand dollars a pop for the pleasure of his telling people to reduce their carbon footprint. Climate change skeptics, with major funding by fossil fuels–supported right-wing think tanks, made it seem like the science was debatable.[32] They talked about theories of sunspots and solar radiation, false data, and parts of the Antarctic that were getting colder. At a 2010 conference organized by the Heartland Institute, a free market think tank skeptical of climate change, attendees were even handed miniature hockey sticks as a symbol of their alternate interpretations of the hockey stick graphs popularized in Gore's film.[33]

So even as *An Inconvenient Truth* helped raise awareness and concern among liberals like me, it had the opposite effect among many conservatives, particularly in the United States. In 1997, 52 percent of Democrats and 48 percent of Republicans agreed that the effects of global warming had already begun, but by 2008, while the percentage of Democrats who agreed that global warming was occurring increased by 24 points, to 76 percent, for Republicans that percentage decreased by 6 points, to 42 percent.[34] Some scholars referred to this change as the "boomerang effect," which is said to occur when a message designed to persuade has the opposite impact of that intended by the communicator.[35] In other words, rather than convincing conservative-leaning Americans that climate change was real and happening, the film and similar efforts to get the word out contributed to an increasing perception that the news about global warming was exaggerated.[36]

Given this political climate, perhaps Gore would not be thought to be the best messenger to speak with a Tangier resident about climate change, yet that was precisely the setup when Eskridge was invited to ask a question of Gore at the televised town hall in the summer of 2017. As the mayor of Tangier, Eskridge could have used the opportunity to ask why the government wasn't investing in communities such as his or to ask for a commitment from climate change activists such as Gore to raise awareness about Tangier Island's plight. But tellingly, Eskridge had a different question. After explaining his background as a crabber with fifty years of experience on the water, he said, "I'm not a scientist, but I'm a keen observer. And if sea level rise is occurring, why am I not seeing signs of it? I mean our island is disappearing, but it's because of erosion and not sea level rise and unless we get a sea wall we will lose our island. But back to the question, why am I not seeing signs of the sea level rise?"

Gore, perhaps assuming that Eskridge was just trying to give him a hard time, took the bait. He asked the mayor, pointedly, to what Eskridge attributed the disappearance of the island. When Eskridge replied that it was due to erosion caused by wave action and storms, Gore then asked, rather skeptically, "So you're losing the island even though the waves haven't increased?"

Yes, said Eskridge. He explained that erosion had been a constant part of life on Tangier ever since its formation. "If I see sea level rise occurring, I'll shout it from the house top," he said. "But I'm just not seeing it."[37]

Watching the exchange as an outsider, it is striking how the two men seemed almost to be talking about completely different things. I brought that up in an interview with author Swift, who lamented the lost opportunity for Gore to explain the difficulty of using anecdotal observation to perceive something as slow moving as sea level rise. As Swift told me, "The biggest problem in terms of the Tangier view assimilating with that of science is that a Tangier waterman has a different way of collecting data. Instead of the model favored by science, a Tangierman goes out on his boat every day and looks at the water. And that anecdotal style of data collection leads him to a completely different set of conclusions. Namely, he's trying to gauge incremental change over the course of decades from the pitching deck of a tiny boat offshore."[38]

But just as Eskridge's limited perspective made it impossible for him to *see* the sea rising up to swallow his island, neither could Gore *see* that the man in front of him was asking him a valid question. Thus, instead of attempting to answer it, he told a parable about a man who, stranded in a flood, asked for the Lord to save him. As the flood waters rise, the man continuously rejects help from others, saying that "the Lord will provide." The man ends up drowning, and Gore concludes: "And I think that we have heaven sent, so to speak, enough solar energy in one hour to provide what the entire world uses for a full year. And from wind, we get 40 times as much energy as the entire world needs. We have the tools available now to solve this crisis. And whether you attribute what's happening to Tangier to what the scientists say it's due to or not, I'm assuming that if you could get cheaper electricity from the sun and the wind, that would be a pretty good deal for you, right?"[39]

Yes, agreed Eskridge, and the town hall moved on to other matters. But the exchange was unsatisfactory at best. When Eskridge asked why he couldn't "see" climate change, what kind of answer was he looking for? And although it was clear from the parable that Gore wanted Eskridge to think differently, what, specifically, did he want him to do differently? The average Tangier resident uses less energy, lives in a smaller home, and drives far less than the average American.[40] Perhaps the approximately 220 voting-age adults living on Tangier are not on Team Climate, but they also represent less than 0.00003 percent of Virginia's population, thus making any attitudes of the islanders unlikely to determine the fate of any future energy policy in the state, let alone nationally. Furthermore, pointing out that green energy will lead to cheaper electric bills might not be very relevant to a community whose home is threatened with disappearance.

From "Please Just Listen to Me" to Something Else?

In 2007, Gore, along with the Intergovernmental Panel on Climate Change, was awarded the Nobel Peace Prize for his efforts to spread the science of climate change, the first time the prestigious award had ever been focused on science communication. In the words of the committee, Gore was "the great communicator," "the single individual who has done most to rouse the public and the governments that action had to be taken to meet the climate challenge."[41] For many, he continues to be viewed as a central messenger for the environmental movement, even making guest appearances on popular television shows such as *30 Rock*, where he played humorously exaggerated versions of himself as a heroic environmentalist running off to save animals in trouble.

But for others, the fact-based appeals to address climate change have fallen short. In 2020, climate change and environmental issues were the two most polarized issues in the United States, surpassing traditional fighting fodder such as immigration, health care, and guns.[42] Although more than 75 percent of Democrats considered climate change to be a top priority, less than one-fourth of Republicans agreed.[43] And as climate change has become more associated with liberal politics, bipartisan efforts to address it in policy arenas have dried up. Some conservative lawmakers who became concerned about climate change pushed back and suffered mightily as a result. For example, Representative Bob Inglis, a Republican from South Carolina, set out to convince his party and constituents that the science of climate change was real, only to lose his seat in 2010 in a landslide defeat. Two years later, another Republican policy maker, Senator Richard G. Lugar of Indiana, followed suit by losing his seventh term to a challenger who called climate change "junk science." These defeats were widely discussed in the media and sent a clear message to other Republican policy makers that climate change was not an issue that resonated with their voters, and if they wanted to get elected (or reelected), one had best steer clear.

This polarization has led to political impasses for getting much done in terms of climate action, not just in the United States, but internationally. As a result, the amount of carbon in the atmosphere continues to increase, and the global temperature continues to rise. In 2006, the concentration of carbon dioxide in the atmosphere was at about 380 parts per million, whereas in 2024, we were exceeding 422 parts per million.[44] And although the latest round of negotiations at the UN Climate Change Conference provided avenues for the countries of the world to transition away from fossil fuels, the agreement is nonbinding and lacks the funding and mechanisms necessary to implement change.[45]

It is this lack of political will and public concern that *Don't Look Up* sought to spotlight, made clear at the end of the film when Dr. Mindy loses his patience and goes on a televised rant. "We should have deflected this comet when we had the fucking chance!" he cries. "But we didn't do it! I don't know why we didn't do it."[46] In the film, it was too late to change course. The "Please Just Listen to Me" model of science communication failed, big time, just as it is not getting things done in real life.

Is this the end of the story? Should we all just throw up our hands and focus on spending time with our loved ones, eating nice meals and comforting ourselves with the thought that at least we tried to get the word out?

We could. But that would be a bit of a cop-out. Spreading science-based attitudes and behaviors is much more complex, and much more interesting, than the "Please Just Listen to Me" approach would lead us to believe. To explore this idea, let's turn to India, which up until about fifteen years ago was one of the last bastions in the world for polio.

CHAPTER 2

Will I Please Just Listen to You?

Polio is a highly infectious viral disease that primarily affects children, has existed for thousands of years, and, at its peak in the 1950s, killed approximately a half million people worldwide each year and paralyzed millions more. Eradicated across most of the world by the turn of the twenty-first century, it still existed in pockets throughout the Global South, most notably in India. In the 1980s, more than 200,000 children around the world were paralyzed each year from polio, and as recently as 2009, India had approximately half of the world's cases and was considered the polio epicenter of the world.[1] Many believed that India—due to a multitude of biomedical obstacles, programmatic deficiencies, and vaccine hesitancy, which was concentrated among hard-to-reach groups in underserved parts of the country—would never be completely rid of the disease.[2] For some, the idea of a polio-free India seemed like an "impossible dream."[3]

But incredibly, the last case of polio was identified in India in 2011, and by 2014, the country was certified as polio-free. In other words, India went from having half of the world's cases to just one case *in under two years*. The story of how India achieved such a resounding success offers lessons not just for other vaccination campaigns, but also for how to spread other types of science-based attitudes and behaviors. It is one of coordination and collaboration, and, perhaps most crucially, it is one where the "Please Just Listen to Me" approach was replaced by something else entirely.

Although the polio vaccine has existed since 1953 and individual countries initiated broad vaccination efforts, not until 1988 did the World Health Assembly pass a resolution to eradicate polio worldwide.[4] In 1995, the Indian government launched a nationwide polio immunization campaign, holding national immunization days to vaccinate young children across the country. Such efforts worked well in many places, but it quickly became clear that millions of children were being missed. To increase uptake, the government began sending vaccinators door to door in regions with low vaccination rates. But unexpectedly, the effort seemed to backfire. As described by Roma Solomon, retired director of the Core Group Polio Project, a global consortium of nongovernmental organizations dedicated to the polio campaign: "Parents started shutting their doors on the vaccinators, refusing to allow their children to be vaccinated, and the enthusiasm of parents turned to reluctance in some states, and strong aggressive resistance in others. The government was faced with a new challenge: understanding why communities would turn against the very people who were trying to protect their children from a deadly disease that had crippled so many children."[5]

Governmental agencies realized that they needed to change their approach, so they engaged volunteers from within vaccine-resistant communities—including nursing students, faith leaders, and teachers—to go on a listening campaign and become ambassadors for the vaccine.[6] These volunteers, mainly women, were trained in interpersonal communication skills such as conflict management and negotiation and tasked with spending time with parents, building trust and rapport through regular visits. In some cases, it took many conversations before families shared their deepest concerns about the vaccine and other issues.

For example, it soon became clear that in the eyes of many poorer residents, getting a child vaccinated could mean losing one or more days of work if the vaccination caused side effects such as fever.[7] In addition, polio was only one of many health concerns. Community members expressed frustration over the lack of other health services in their communities and expressed concern that their children were dying from other diseases. To address this concern, "health camps" were set up in highly resistant areas, which convened medical personnel and provided free medication and vaccines on days organized in advance with community leaders. This effort was hugely successful, leading to increased trust among the communities and agencies. As stated in one article about the process: "When talking about health care services in urban slums, an interviewee described that, 'it's not a matter of hard-to-reach but rather, hardly reached.' Communities felt ignored by their government, and were thus mistrusting and skeptical of government or NGO intervention during polio vaccination rounds."[8]

For example, the initial visits from unknown vaccinators had been viewed as threatening in some regions because some viewed the vaccine as connected to coercive measures by the state to control family planning. Another aspect of resistance was due to religious considerations because in some regions the vaccine was increasingly seen as forbidden by Islamic law. To address this concern, health workers sought support from religious leaders, and regular conversations and connections led to increasing announcements during sermons about the safety and importance of the vaccine for children. And although popular Bollywood stars helped spread the message of the vaccine nationally, local press and media were even more important for communicating with "hardly reached" populations in target areas.

India's polio vaccination campaign took extraordinary efforts, but the payoff was even more outstanding. By 2024, thirteen years had passed since the last confirmed case of polio in India.[9] Perhaps most notable about this achievement is that these interventions had a spillover effect, leading, in some high-risk regions of the country, to an increase in full routine immunization coverage. For example, in Uttar Pradesh, full coverage increased from 36 percent in 2009 to 81 percent in 2016.[10] India's mobilization of civil society during the polio immunization campaign also contributed to the success of the COVID-19 vaccination campaign, as the same community mobilizers who met repeatedly with parents about polio were called up when COVID-19 first hit.[11]

Understanding Vaccine Hesitancy

Vaccines have been around for hundreds of years, and for as long as they have been around, so has vaccine resistance.[12] Vaccines can feel scary and counterintuitive. The science behind them is complex and sometimes controversial, as seen in the case of COVID-19 and the novel RNA technologies. Vaccines typically need to be injected into otherwise healthy people—often small children or infants. And some vaccines inoculate against diseases that most of us would likely never get. In that sense, getting a vaccine, or having your small child get a vaccine, is more of a public service act than a personal choice for one's own health. For all these matters and more, vaccines would naturally be a tough sell for science communication.

Even so, vaccination is one of the most effective public health tools we currently have, one of the best science-based behavioral and societal tools. Vaccines can significantly improve not only the lives of those who get them, but also their family members, friends, and neighbors, as the more people are vaccinated, the less chance the virus or the bacteria will be carried or transmitted.[13] And vaccines are widely accepted across different cultures, countries, and religions. For example, as of 2022, global coverage with three doses of hepatitis B vaccine was estimated at 84 percent, and 83 percent of children received one dose of measles-containing vaccine by their second birthday.[14]

Conversely, when vaccination is not widespread, decreased public health outcomes can result. Some diseases, such as measles and pertussis, require upward of 95 percent of the population to be vaccinated to create "herd" or "community" immunity, which is reached when a large enough proportion of a population becomes immune through previous infections or vaccination and thus cannot spread the disease.[15] This level of public acceptance can be incredibly difficult to achieve, and even after the threshold is reached, can slide backward if something happens to reduce vaccination rates. For example, in 2022, a combination of reduced access to and increased hesitancy toward childhood vaccinations in Zimbabwe resulted in a measles outbreak that killed more than 700 children and brought serious illness to thousands of others.[16]

Because attitudes and behaviors toward vaccines matter not only for one's personal health but for the health of others, they are akin to other key science-society issues in that it matters if not everyone is onboard. As such, researchers across multiple fields have long sought to understand what drives vaccine hesitancy and what approaches there are for addressing it. Interestingly, it is perhaps what they have *not* found that is at least as interesting as what they have found.

The first thing researchers have not found is a black-and-white binary between the vaccinated and the unvaccinated. Rather, hesitancy exists on a spectrum. Some people may absolutely refuse to get all vaccines, but they are in the minority.[17] Most hesitancy stems from concerns about the safety and effectiveness about a particular vaccine and is often about postponing vaccination rather than absolute refusal.[18] To be honest, I am somewhere along this spectrum with the flu vaccine, which I only get rather reluctantly and almost always at the urging of my mother.

Another thing researchers have not found are consistent sociocultural patterns in vaccine acceptance. That may seem counterintuitive, as one thing we all *think* we know about vaccine skepticism, particularly in a post-COVID world, is that it is associated with religious beliefs, political ideologies, and public ignorance of science. But research has found that the main determinants of vaccine hesitancy are far more complicated and context specific.[19]

Yes, for certain vaccines, such as with the case of polio in India, there is a demonstrated relationship between religious beliefs and vaccine hesitancy. But in other cases, religion has not been found to be a significant predictor of vaccination acceptance, and some studies have even found that it can affect vaccination positively.[20] Religious-based vaccine hesitancy or acceptance seems to depend not just on religious affiliation, but on the level of religiosity, the country, and the vaccine in question.[21] For example, hesitancy toward the human papillomavirus, or HPV, vaccine has been found to be linked to religious beliefs in some places, but not in others.[22] Furthermore, even when a relationship is found between religious beliefs and vaccination rates, it often is less a matter of causation and more of one of correlation. There is evidence that in many cases, supposedly religious reasons to decline immunization are not really about faith-based objections per se, but typically reflect other social and health-based concerns.[23]

Another common assumption is that political beliefs play a strong role in vaccine hesitancy. But again, this assumption is largely context dependent. Taking COVID-19 as an example, conservatives in the United States were less likely to get vaccinated against COVID-19 than liberals, but the opposite pattern was found in the United Kingdom, where conservative voters were more likely than Labour party voters (the political left) to get vaccinated.[24] Such trends point to the larger role of partisanship, rather than politically related ideological beliefs, as a key driver of skepticism. For example, surveys of more than twenty thousand people across nineteen countries found that a key determinant in whether people agreed that their government was handling the coronavirus outbreak well was whether the party they supported was the one "calling the shots."[25]

Public ignorance of science is another factor that is often thought to be associated with vaccine hesitancy or refusal. But again, the research paints a more complicated picture. Although some studies have found a relationship between lower levels of education and vaccination intention, others have found education level not to be an accurate predictor.[26] Indeed, some studies have shown the opposite trend—higher levels of skepticism have been associated with increased access to and levels of formal education.[27] Interestingly, research suggests that parents who opt into vaccines typically have little knowledge about vaccination as compared to noncompliant parents, who often have greater interest and have spent more time looking up information on vaccines and other health issues.[28] These studies have found that what drives parents' choices is often just "following the doctor's orders" rather than any increased knowledge about a particular vaccine.

Research that has attempted to find patterns across different contexts, vaccines, and places has mostly come up short. The main finding has been that vaccine hesitancy is incredibly complex, with diverse reasons for people choosing not to vaccinate.[29] Thus, although vaccination is often framed as an epidemiological task—a matter of prescribing a one-size-fits-all approach to the masses—whether people decide to get vaccinated often comes down to very individual reasons and contexts.[30]

However, underlying this diversity of factors is the issue of trust.[31] Trust can mean many things—trust in institutions, in modern medicine, in individual health care workers, in science more generally.[32] Most of us have varying degrees of trust in all the above. Few of us would say we are 100 percent trusting of government, of pharmaceutical companies, or of the science of vaccinology. But deciding to put a needle into our arm for a vaccination takes a leap of faith—it says that we have enough trust in scientific experts and institutions and that doing so is more likely to bring us benefit than to cause us harm.[33] As sociologist Barbara Misztal wrote, "To trust is to believe despite uncertainty."[34]

Reaching Out to the Vaccine Hesitant

If trust is so central to vaccine acceptance, what does that mean in terms of what does, and doesn't, work for increasing vaccine uptake?

First let's talk about what doesn't work. One approach that is increasingly common, yet often counterproductive, is shaming. I think most of us know that shaming doesn't work, yet we do it anyway and are often encouraged to do so by our political leaders. For example, Tony Blair, former prime minister of the United Kingdom, said during the COVID-19 pandemic, "Frankly if you are unvaccinated at the moment and you're eligible and have no health reason for being unvaccinated, you're not only irresponsible but you're an idiot. I am sorry but truthfully you are."[35]

Shaming doesn't typically motivate prosocial behaviors and can have the negative side effect of leading a person to withdraw socially.[36] Furthermore, shaming can backfire in a big way, as people will instinctively look for support for their beliefs elsewhere, often going deeper into antivax communities and rhetoric.[37] That's partly why vaccine resistance tends to occur in social and geographical pockets, where skepticism is reinforced by others with similar beliefs in one's social network. For example, a study of childhood vaccine exemptions in California found that vaccine-hesitant clusters tended to occur in predominantly White and affluent areas such as Beverly Hills and Santa Monica as opposed to neighboring low-income and more racially diverse neighborhoods.[38] These pockets of homogeneity enhance the perception that vaccine refusal is both a safe and socially acceptable decision, leading to a reinforcing mechanism where individual preferences are reinforced rather than challenged, and attracting other parents inclined to opt out of vaccines. In other words, if we're part of a community where vaccine hesitancy is common, we're less likely to be vaccinated.[39]

One of my favorite comedians, Aziz Ansari, pointed out the utility of shaming in a short Netflix special, *Nightclub Comedian*, in which he took aim not just at vaccine skepticism, but of those who ridiculed people for their skeptical beliefs. Ansari had recently lost his uncle, who had refused to get the vaccine, to COVID-19, and his family was heartbroken at the loss. But Ansari wanted the (mostly vaccinated) audience to know that shaming people for not getting the vaccine wasn't helping matters. He said, "I don't think . . . any of these people are idiots. . . . I just think they're trapped in a different algorithm than you are. . . . And if you're calling them idiots and stuff, you're trapped in another algorithm. . . . We just gotta figure out some way to have some empathy. We're all just kinda trapped in our own little world. And unless we figure out how to talk to each other in real life again, it doesn't matter what the problem is."[40]

Shaming also downplays the very real fear that many vaccine-hesitant people feel about the potential side effects of the vaccine. Communities that have historically been marginalized often have negative views of medical systems due to histories of unethical treatment, contemporary disregard for concerns, and limited access to ongoing and quality health support. For example, patients from Black communities in the United States still experience racism and lower levels of treatment in the health care system compared to White patients, leading to reduced health outcomes.[41] Thus trust in vaccines is harder to come by when people have had bad experiences with health care institutions in the past. For example, Solomon of the Core Group Polio Project explained that one major determinant of whether people were suspicious of the polio vaccine was whether they lived in a region where access to, and the quality of, health care services was poor.[42]

Another approach that hasn't been shown to be very effective, to the frustration of many public health officials, is providing factual literature (for example, leaflets or websites) that tackle common vaccine-related misconceptions (such as the erroneous belief that vaccines cause autism).[43] This approach is largely based on the assumption that if vaccine-hesitant individuals had the correct information, they would change their minds. But as explained in chapter 1, humans are not Spock-like creatures who interpret facts on a neutral playing field. Although debunking misinformation can be a useful strategy in many cases, it has limited effectiveness among vaccine-hesitant individuals. For example, one study tested four types of messages commonly used by public health agencies to promote acceptance of the measles-mumps-rubella vaccine among parents.[44] The researchers surveyed parents' beliefs and attitudes about vaccines and found that none of the messages increased intent to vaccinate, and in some cases, the interventions backfired.[45] Compared to the control group, who received no information, parents who were exposed to the messages were less likely to vaccinate their children, and that was especially true among parents who had already expressed some degree of hesitancy. Further, the messages designed to evoke concern (for example, a story about an infant who almost died of measles) led to greater misunderstandings and deeper concerns about possible side effects of the vaccine.

And no, catchy slogans, such as "Don't Hesitate, Vaccinate!" don't do wonders for convincing the vaccine hesitant either.

So what does work?

I was humbled to learn that one of the biggest drivers of vaccine acceptance is conformity, driven by social norms within one's network.[46] In other words, we make decisions about whether to get vaccines not because we're smarter, more "scientific," or more righteous than those who don't, but rather because we're part of a community where those around us are also getting vaccinated. That's why doctors are encouraged to use "presumptive" language when speaking with parents about vaccinations (for example, saying "so she's due for three vaccines today" and not "are you planning to vaccinate your child?"), thus communicating that vaccines are a social norm.[47]

We like to think of ourselves as independent, logical thinkers, but the truth is, we're hugely influenced by what those around us are doing and thinking (and what they think about what we're doing and thinking). In other words, we're often just going along with the crowd. And I would argue that—in the case of vaccines, as well as many other science-based behaviors—that is a good thing. Because of our miserly brains, which are prone to bias and suboptimal decision-making, we *should* go along with the crowd if that crowd is following the advice of experts in matters in which we have little expertise. But we need to be clear that most of us don't personally *know* that a given vaccine is safe, or effective, or not going to implant us with microchips. We simply are part of a community that trusts the medical and scientific establishment that tells us it is so. Once we understand that we're not talking about critical thinking, logic, or even facts, but rather about trust, social norms, and experiences, what we mean when we talk about the goals of science communication and evidence-based behaviors changes. It stops being about a war of who knows more and more about creating a sense of access to, and trust in, the benefits of science.

How can we increase the uptake of science-based ideas and behaviors in communities where a significant number of people are inclined to mistrust, and thus reinforce social mistrust, in those same ideas? To answer this question, we need to understand a bit about the science of how behavior spreads.

From Simple to Complex Contagions

We often think about ideas and behaviors—whether we're talking about the uptake of vaccines or social behaviors such as recycling—as spreading in the way a virus does. One person becomes exposed to the new idea or behavior and then passes it on, in the same way as sitting next to someone who has a cold on a long train ride might lead to (inadvertently) sharing their germs. This idea of viral contagions is seen in the way we talk about the power of social media, particularly platforms such as YouTube and Instagram. A video or meme is said to "go viral," which means that people share it with lightning speed, so something can go from almost no views to millions of views in just hours.

But whereas simple information, such as job advertisements, can spread like viruses, researchers have found that such spreading does not apply to complex ideas or calls for behavioral or societal change. This type of more complex information will come up against the biases and worldview hurdles discussed in chapter 1. If someone shares something that is geared toward changing our minds or behaviors about something of importance, there will be a degree of resistance that will need to be addressed. In his book *How Behavior Spreads: The Science of Complex Contagions*, sociologist Damon Centola called these scenarios "complex contagions" and argues that they are unlike "simple contagions" in which increasing amounts of factual information is spread virally across a population. He wrote: "This basic problem of diffusion—that is, the failure to spread behavior—occurs whenever behavior change encounters resistance. Attempts to spread everything from vaccinations to innovative technologies to environmentally friendly business practices have faced similar difficulties. The less familiar an innovation is, and the more inconvenient, uncomfortable, or expensive it is, the greater the resistance will typically be, and the less likely it will be to diffuse."[48]

Thus messages that are spread broadly can lead to backlash if they are perceived to deviate from norms and values among one's social group (think of how *An Inconvenient Truth* was received by conservative voters in the United States). The broad uptake of new ideas requires support within others in a social network. Complex contagions therefore require something more powerful—social confirmation. And according to Centola and others, social confirmation typically requires two intersecting ingredients.

The first ingredient is direct, personal interactions with someone who has already taken on the new belief or behavior. In a community where a behavior has not yet been adopted (or where it is being actively resisted), initiating such interactions poses a challenge. For example, if one's family and friends are not willing to get a vaccine, it is hard to know where a direct interaction with someone who has gotten the vaccine might emerge. But this dilemma poses an important opportunity, one that health care workers in tight-knit communities have long known: if we want to give new behaviors credibility, we need people to see people they know and trust taking on those behaviors. Those promoting the vaccine must be counted as among those who are known and who can be trusted.

Let's revisit the polio example. One reason the campaign was so successful was because it was spread by trusted community members and health workers who spent time building trust with parents and addressing their concerns about the vaccine. For example, Dr. Naveen Thacker was a young pediatrician who became a key voice for the polio vaccine in Gujarat, one of the epicenters of the polio epidemic in the 1990s. Thacker wrote about how his personal connections with individual patients was not only what helped increase acceptance of the vaccine, but also what helped him persevere.[49] One of the most challenging people he had to persuade was a father in a very isolated area of the region. The father said that he would only consent to allow his children to be vaccinated if Thacker made a commitment to looking after the health of his family for the rest of their lives. The doctor made the deal, which was the beginning of a multigenerational relationship—one grounded in trust and friendship—between him and the family.

Although few health professionals would have the capacity to make a lifetime commitment to families, the story points to the importance of a personal and direct connection through which trust in the polio vaccine was established. Countless studies of what works in changing the minds of vaccine-hesitant individuals have consistently found that a crucial factor was a reassuring conversation with a trusted health care provider.[50] Thus social confirmation can come from professionals—if those professionals take the time to demonstrate care and compassion to hesitant publics.

The second ingredient needed for social confirmation is redundancy.[51] Redundancy is often associated with something that is not useful because it already exists (for example, when someone is "made redundant" at a workplace). But redundancy is essential in complex contagions because social confirmation requires someone hearing the same (or similar) message from different sources to validate it as worth listening to. Studies have found that the more individuals from one's own network who adopt a given innovation or belief, the more likely one is to follow suit.[52] The riskier the idea or innovation, the more redundancy is needed in one's social network for a person to adopt a new behavior.

Thus, in vaccine-resistant communities, redundancy means that a lone doctor is unlikely to make a change by themself. But a doctor, a teacher, or an epidemiologist who happens to live down the street and brings over soup when a child falls sick—that person might provide enough redundancy to spark a change of mind in one or more people within the community. In this sense, redundancy provides a positive feedback loop: more people who adopt the behavior in a given community increase the chances that others will also. When there are enough adopters in a given community, they can become the critical mass that shifts the tide, triggering "social tipping points" in social norms and leading to drastic changes in popular opinion.[53] That's what happened in India: as more and more parents in vaccine-hesitant areas began vaccinating their children, it became a social norm, and people went from hiding their children when vaccinators came knocking to bringing their children willingly to vaccination days.[54]

Because of the positive feedback loop inherent in redundancy, social norms can sometimes seem to change quickly. For example, one well-cited example of the importance of social confirmation for social change was the fight for LGBTQ+ rights in the United States. Although members of the LGBTQ+ community have advocated for equal protection under the law for at least a century, only since the early 2000s were many of these rights enshrined into law and practice. For many people, the shift in public perception seemed to happen overnight. In 2004, only about one-third of people in the United States were in favor of same-sex marriage, but within a decade, that number had jumped to more than half—a critical mass that led to the legalization of same-sex marriage in 2015.[55]

Scholars largely attributed the shift to people being influenced by others in their social network—specifically with others with whom they shared strong ties. During the 1980s and 1990s, more and more individuals "came out of the closet," and people who had previously been against gay marriage, among other equal rights for the LGBTQ+ community, suddenly found that their family members, friends, and coworkers were being negatively impacted by those beliefs and related policies. Indeed, surveys have found that the main reason people cited for changing their minds about same-sex marriage is because they knew someone who was gay.[56] The more people who supported same-sex marriage, the more that doing so became a social norm. The synergistic mix of strong ties and redundancy changed the conversation and, ultimately, the law.

An Army of Trusted Voices

The success of India's polio eradication campaign offers several lessons that matter not only for other public health campaigns, but for thinking about how to spread information about science and increase adoption of science-based attitudes and behaviors more broadly.

First, it is important to emphasize the importance of efforts that target specific rather than broad audiences for maximum impact. The polio campaign built off this strategy, thus ensuring that the hardest-hit areas received the most resources, and used the power of closely knit networks to encourage social acceptance of something that otherwise might have been met with resistance. Similarly, messaging from trusted local media and newsletters was seen as more important than national campaigns driven by celebrities, which could get reach but not depth. This lesson points to the importance of going beyond a broad-brush approach—it means identifying specific communities and groups and physically going to and engaging with trusted individuals in those places.

Relatedly, building relationships of mutual understanding and open communication is key. Public health officials in India worked closely with civil society, connecting with crucial community and religious leaders in vaccine-resistant regions of the country. Adopting an approach in which dialogue went back and forth and health officials listened to concerned parents was crucial to this effort. And seeing vaccination as part of a broader strategy to increase health and well-being in "hardly reached" communities ensured that the campaign paid additional dividends. In this sense, building trust in the vaccine was a collective and coordinated effort that was dependent on an army of trusted voices.

British sociologist Anthony Giddens once referred to personal interactions between experts, such as doctors or scientists, and laypeople as "access points," which he deemed essential for maintaining trust in modern institutions.[57] Beyond just providing information, Giddens argued that such interactions provide opportunities for experts to share values and character traits—warmth, integrity, professionalism—essential for building trust. This concept points to an opportunity for members of the scientific community—whether researchers, educators, or communicators—to serve as "access points" by connecting, in a direct and personal way, with people from communities that have been underserved by science. Indeed, in some ways, those of us who work in the research community are in an ideal position to play such a role. Here I will share some good news. It stands to reason that of the top three most trusted professions in the world, number one is doctors and number three is teachers. Both professions enjoy regular interactions with members of the public, whether through annual checkups or parent-teacher conferences. But what is the number two most highly trusted profession?

It may come as a surprise to many of us, but the answer is scientists.[58] If you don't believe me—and I would understand if you didn't—take a moment to search for the data for yourself. Although this trend varies somewhat from country to country, both medical and research scientists are consistently ranked to be among the most trusted professions in research polls across different cultures and communities.[59] However, there are some important caveats. First, such trust is somewhat moderate rather than strong. For example, in the United States, more than twice as many people trust scientists "a fair amount" to be acting in best interest of the public than who trust them than "a great deal."[60] Furthermore, trust in scientists, and the perception that science is a positive force for society, is decreasing in some places. For example, one survey conducted in United States found that whereas 86 percent said they had had confidence in scientists to act in the public's best interest in December 2018, only 69 percent reported the same in May 2023.[61] Surveys of adults in the United Kingdom found that while in 2002, 87 percent agreed that science has a positive impact on society, the number declined to 79 percent in 2022.[62] Globally, a survey conducted in 2023 reported that more than half of those queried believe that science has become politicized in their countries.[63]

Rather than a general lack of understanding or trust held by society, research has found that public perceptions of scientists and scientific information are much more nuanced. Research suggests that more than trying to just share facts, scientists can increase trust if they are perceived as honest, caring, and competent.[64] Additional qualitative research backs this conclusion up by finding that scientists can be extremely effective messengers—but their messages need to be firsthand and personal to be of greatest value. In other words, what works for doctors in terms of having a good bedside manner can also work for scientists.

Here's where we hit a snag, though. Scientists and researchers do not typically have the type of face-to-face interactions with the public that family physicians regularly enjoy. We do not see science deniers for their annual cognitive checkup (unless you count that wine-fueled debate about nanoparticles with your uncle at Christmas dinner, which you really shouldn't). Researchers might talk with the media, but in such encounters, it is a reporter, and not us, asking questions or listening. And when we do have conversations about our research, it is most frequently with other scientists, far from public arenas.

Furthermore, not all contact between scientists and the so-called public leads to good outcomes. Because of the prevalence of the "Please Just Listen to Me" model, many scientists' attempts to present the facts in a straightforward manner can fall flat or even be counterproductive if not done in a thoughtful manner.

So how can scientists become access points in the way Giddens describes, and how can doing so lead to outcomes for science to be impactful for policy, practice, and society? That's what I set out to explore when, in 2012, I boarded a small plane to head to the Madidi region of Bolivia for the first time.

CHAPTER 3

From Impact to Encounter

FLYING NORTH FROM BOLIVIA'S CAPITAL CITY of La Paz into the depths of the Amazon rainforest is quite a journey. One begins at El Alto airport, situated at the crisp, cold altitude of four thousand meters above sea level, and boards a ten-seater plane that gets beaten about by high winds until it reaches the mountains of the Cordillera Real, one of the ranges of the Bolivian Andes. The plane cuts a path between the peaks, so looking out the window one can see homes, people, and animals dotted across the range, providing an intimate glance into a world that would otherwise take days of arduous travel. But almost as quickly as the mountains appear, they fade into the distance as the plane passes the range and begins to cross Las Yungas, the subtropical region known for its complex mosaic of lowland rain and cloud forests. Subtropics turn to full tropics as the plane begins its final decent into the thick canopy of Bolivia's Amazon Basin, brown rivers twisting like snakes through the deep green of the forest.

Although there is a way to travel by land, it is an arduous and exhausting journey along one of the most dangerous roads in the world. The public bus takes eighteen hours, with one pit stop in the wee hours of the morning, and private transportation is not much faster and even more of a heart-in-your-throat experience. About halfway along the route, traffic suddenly switches lanes with the downhill drivers now driving on the left side of the road, supposedly to make passing safer. But on a road with few guardrails and even fewer signs, a driver must generally hope for the best—that there are no tourists who haven't gotten the memo to drive on the opposite side, that the road isn't washed out, and that the heavy trucks bringing goods from the city to the countryside have working brakes.

Whether one travels by plane or overland, the destination is worth it. The Madidi region, as it often called, is one of the most biodiverse on the planet. It is home to 11 percent of the world's bird species and an estimated twelve thousand plant species, as well as many endangered mammals such as the white-lipped peccary, vicuña, Andean bear, and giant otter, among others.[1] It encompasses five life zones in which seven Indigenous groups (Takana, Tsimané, Moseten, Takana-Quechua, Lecos, Quechua, and Esse-Ejja) reside, along with more recently arrived highland communities of Aymara and Quechua descent. In addition, it is at risk from local and global threats to biodiversity such as deforestation, climate change, and mining.[2]

As such, the region is a beacon for scientists of all types from all over the world. On the flight to Madidi, I would frequently run into other researchers; some were visiting the region for the first time, others for the fiftieth. At the local café, I would surreptitiously look around to see foreigners and Bolivian urbanites working on their laptops and entering data into Excel spreadsheets. It was a running joke among the Madidi park staff that one could throw a stone in any direction and hit a scientist.

I, too, was drawn to the incredible biological and cultural diversity of the region. Unlike other researchers, though, I wasn't studying the plants or the animals or even the people. Instead, I was doing something that might sound a bit strange to some: I was "researching research."

Researching research. It was a topic that produced chuckles and raised eyebrows whenever it was mentioned. "Well, I guess the scientists have nothing left to study," was the reaction of one man who asked me what my PhD was about.

But in a place like Madidi, to "research research" meant to study the impact of an activity with a fraught and complicated past, present, and future. Madidi's history of scientific research and exploration began with the arrival of the Spanish colonizers and their discovery and exploitation of the region's natural resources—quinine and rubber—and even a search for the fabled city of El Dorado.[3] Under colonial rule and after independence, outsiders were drawn to the region, seeking adventure, riches, and, increasingly, the answers to scientific questions. Madidi was a place where there were still many species of plants and animals that had not yet been "discovered" and where, every few years, reports of uncontacted peoples attracted anthropological expeditions into the depths of the jungle.[4] It was a place where research was a known activity, if a contentious one, and where the locals had their own vernacular to describe this ubiquitous, mysterious endeavor that was alternately seen as vitally important or completely useless, depending on who you spoke to.

Indeed, one of the first new words I learned in Madidi was "*tesista*," which translates literally to "thesis maker." I, apparently, was a "thesis maker," following in the footsteps of countless others who had come to the region to conduct their research and build their careers. Most times, people used this term rather neutrally, but I would soon learn that it came with baggage I didn't realize I was carrying and I would not be able to leave behind. "So many researchers," said a leader of the Takana Indigenous council, after I introduced myself. Indigenous educator and scholar Linda Tuhiwai Smith writes of the legacies of western research on Indigenous lands, where (often well-meaning) foreigners travel to "exotic" places to study them for the purposes of science. Of the term *research*, she wrote that it is "probably one of the dirtiest words in the Indigenous world's vocabulary. When mentioned in many Indigenous contexts, it stirs up silence, it conjures up bad memories, it raises a smile that is knowing and distrustful."[5]

At the same time, however, research could not be seen without its potential, whether real or imagined, of some fruitful or lucrative outcome. After all, in Madidi scientific research had led to the creation of the parks in the first place, when biologists had helped draw the boundaries of the protected areas in the 1990s.[6] Scientific research also provided important information about threats, such as climate change and deforestation, to the human and nonhuman inhabitants of the region.[7] And the "discovery" of the region as one of the most biodiverse on Earth drew not only scientists, but tourists, who flocked to the many locally owned ecotourism lodges in ever-increasing numbers.

So "researching research" in Madidi was politically and socially complex, with many possible and nuanced interpretations. Yet it was also an important topic for my PhD, in large part because researchers in my academic field—conservation science—were coming to the realization that despite decades of brilliant and compelling scientific research, we have largely failed in our mission of preserving and protecting the planet's flora and fauna. Although our science was increasingly effective at documenting threats to biodiversity and ecosystems, we seemed no closer to helping develop and implement the solutions to save them.[8]

Researching Research

Conservation science (or conservation biology, as it was traditionally called) is perhaps a unique field of science in that it is supposed to be guided by its goal of providing the principles and tools for preserving biological diversity rather than its contributions to scientific knowledge.[9] It was first established in the 1980s as an openly "mission-oriented, crisis discipline," laden with the promise of representing a new kind of science that would seek to engage directly with the so-called real world. David Ehrenfeld, founding editor of the journal *Conservation Biology*, wrote of its history: "Its mission is to save as much as possible of the earth's biodiversity. We consider ourselves physicians to nature. . . . Developing the frontiers of knowledge by contributing to the world's theoretical and practical understanding of biology is exciting and is necessary for tenure and promotion for those of us who are academics, but our real mission is to save the life of the planet."[10]

However, its founders were clear that such potential had to come with a warning, cautioning that the new "mission-driven" discipline could also fall into the typical "business as usual, blinders in place" tendencies of academia if its scientists did not make explicit efforts to act differently. "If conservation biology becomes isolated in the mental world of academia," wrote Michael Soulé, one of the founders of the discipline, in 1986, "it will be of little use."[11] Thirty years later, when I was finishing my PhD, these origins had evolved into a crisis of purpose in the field.

Such concerns were enmeshed in larger discussions within academia about its responsibilities to society, often framed in terms of research not having "impact." I saw this concern raised particularly in the United Kingdom, where I was doing my PhD and where there was a furious debate about a new requirement for academic institutions to submit case studies that demonstrated "impacts beyond academia" to receive national funding.[12] Back home in the United States, the National Science Foundation was evaluating research projects not only in terms of "intellectual merit," but also in terms of "broader impacts."[13] I would soon learn that similar conversations were happening across the world and across disciplines, researchers, communities, and politicians—all of whom were asking for "research with impact."[14]

Although this call seemed straightforward, when one scratched just a tiny bit beyond the surface, it was anything but. Everyone seemed to want "impact," but there was no consensus on what it was and how it might be recognized, understood, or evaluated. Perhaps even more importantly, there was a lack of clarity on how research could lead to predetermined impacts, as well as increasing concern that unintentional negative consequences of "impact agendas" often went unrecognized.[15]

For example, on one hand, funding agencies and scientists were increasingly discussing how to "bridge the gap" between research and practice, emphasizing the role of scientists to "translate" their research and provide evidence-based solutions to decision makers.[16] In this reading, the problems that society faces are often framed in terms of a lack of clear, persuadable knowledge needed to make the right decisions, and thus it is the job of scientists to fill this "gap" by producing the missing information, an approach that has often been described as the "deficit" model of science communication.[17] But in other published literature, particularly in the social sciences, these types of models were being increasingly critiqued as not being an accurate description of how science is used, or not used, in society. Decades of research from multiple fields had demonstrated that empirical evidence is only one factor (and often a minor one) influencing decision-making and change, as "facts" are not perceived in the same way by different publics, as pointed out in chapter 1.[18] Scholars were concerned that such a linear, trickle-down approach to impact whittled knowledge down to being a "thing" to be transferred between a group of self-selected experts and the rest of society, which failed to recognize that the extent to which research is seen to "have impact" is largely based on who is being asked.[19]

Thus a large part of the critique focused on the lack of inclusion of lay voices in conversations about what constituted impact, particularly those of the so-called users of the research, such as policy makers and practitioners (for example, environmental managers, doctors). In my field of conservation science, it was increasingly recognized that the voices, ideas, perspectives, and concerns of other actors, including farmers, Indigenous communities, park administrators, and heads of households, also needed to be considered in such discussions due to the important role of such groups in determining how land and resources are used and managed.[20] For example, biologists Andrew Balmford and Richard Cowling wrote: "If we want to move beyond documenting losses or identifying specific causes of decline to understanding their underlying drivers and implementing interventions on anything other than a piecemeal basis, we need to undergo what one of us describes as "an epiphany for . . . natural scientists": the realization that conservation is primarily not about biology but about people and the choices they make."[21]

Thus scholars were increasingly realizing that current calls for "research with impact" needed to look more carefully not just at what information was needed and how to get it into the right hands, but about who decides—who is included and who is excluded in such discussions. In other words, *who* is at the gap between research and action, and why do their voices matter?

Impact Is in the Eye of the Beholder[22]

In Madidi, I collaborated with Bolivian biologists and social scientists to better understand diverse perspectives on what constitutes impactful research.[23] We conducted interviews and focus groups with more than one hundred individuals, including community members, Indigenous leaders, park staff, and scientists, asking general questions about peoples' perceptions of scientific research and about the impacts of specific studies. Part of the data collection involved completing a systematic review of past research in the region, for which I combed through offices in protected areas to find research permits and then called, emailed, or met in person with the lead researchers to verify the information obtained and ask additional questions about the potential implications and outcomes of the research.[24]

The first thing to become clear was that it was, indeed, important to talk about the "products" of research. That wasn't because there was clear evidence that such "products" were generating actionable changes in policy and practice, as the deficit approach implied, but rather because their absence was creating a massive problem of distrust and resentment toward research among certain communities. To sum up the most expressed sentiment by local people during this process, "Researchers leave nothing behind."[25]

The evidence largely supported this perception. After completing the systematic review of research, we found that only about a third of researchers attempted to hand over written results to local actors, such as community leaders and protected area administrators, and fewer than 25 percent presented results orally.[26] The most sobering finding of all was that these trends were largely driven by foreign-based (typically coming from North America or Europe), rather than Bolivian-based, researchers. Although there was almost a fifty-fifty split between foreign- and Bolivian-led research projects conducted in the Madidi region, in only one instance did a foreign-based researcher engage in any form of local, regional, or even national knowledge exchange about the findings of their research.[27] These low levels of local dissemination are especially striking given that to be granted permission to do research in the protected area, a researcher must sign a formal agreement committing to share the results of the study with management in that area.[28]

It wasn't that the research in question was without any potential applied value. Most researchers (83 percent) stated in the affirmative that their research had implications for both policy and practice. Even more striking, 70 percent listed potential applications of their research at the local scale (such as contributing to park management plans by providing information of species inventories and other biophysical information).

Perhaps one might assume that the studies in question weren't completed or the results were inconclusive. But that wasn't the case. It wasn't that the foreign-based researchers didn't disseminate the results of their studies; it was just that they didn't do so in Bolivia. Rather, they were far more likely to share results within the global scientific community (more than 70 percent published their findings in international journals or with foreign-based institutions). Similar findings were found in a review of research carried out in Manu National Park in Peru, which found that although North American authors published far more papers than Peruvian authors about the ecology and biodiversity of Manu, these papers were typically found in English language journals. In contrast, Peruvian authors tended to distribute the findings of their research in unpublished reports in Spanish and were more than four times as often as North American authors to offer management recommendations based on the research.[29] This trend of foreign researchers taking the results of research outside national borders has a long history stemming from colonial times, when generations of European explorers and naturalists collected specimens and other data to send back to natural history museums and zoos back home. For example, as late as the 1990s, more than 90 percent of approximately thirty-seven thousand zoological specimens from Bolivia were in collections beyond its borders.[30]

So one piece of the puzzle was startlingly clear: researchers were largely doing research that they thought could have some applied value, but they weren't sharing that knowledge with those who could potentially use it. In this sense, it felt like much research in the region had more of a negative than a positive impact, at least from a local perspective.

As we were coming across these negative impressions of research, another story was emerging. Despite a lot of local frustration and resentment toward researchers, people were not rejecting research projects across the board. Rather, most Indigenous leaders and park managers with whom we spoke said they wanted more research, more scientists. They wanted to know what prior research had been done and had ideas for what could be done in the future. In other words, they saw that research *could have* some value and importance, and the lack of discernable results was a missed opportunity. Indeed, some communities had even gathered resources to pay for it. For example, in the 1990s, leaders of the Takana Indigenous nation raised money to translate a 1950s book written by a German anthropologist about the Takana people into Spanish. In an interview, Celín Quenevo, then president of the Takana people, described his thoughts on the matter:

> I think that research is very positive if it used by the people who really need it and where it comes from. But if information is gathered and then just taken away, and doesn't return to its place of origin, then we do not know what purposes that information can serve. . . . So I say yes, research is important to the extent that the information returns, and it is, at least, in Spanish. Because we have the experience that some research has been done by German anthropologists, they have made several books, all in German. And we had to go through a series of procedures, even outside the country, to ask for the author's permission, until we finally had it translated with the permission of the German embassy. And all this just to discover what they had written![31]

Another finding, one I'd seen expressed in the literature, was that it wasn't just the lack of sharing back results, but that the research being done wasn't the research that was needed for local decision-making. Environmental managers and community leaders were frequently seeking answers to questions researchers hadn't explored, at least not in the region. For example, gold mining was arguably the largest environmental threat in the Madidi region, with dozens of unregulated mining operations both inside and outside the protected area. Community leaders regularly mentioned the need for more information on the environmental and health impacts of gold mining, particularly the mercury used in mining, on the health and environment of the river communities who largely relied on fish for protein. I spent a couple of hours one afternoon with leaders of the Tsimané-Moseténed Indigenous council as we searched for anything written on the issue. But despite the urgent threat, in my database of research projects there was only one study focused on gold mining. In comparison, approximately forty projects within a ten-year period focused on documenting various flora and fauna in the region. I bring this point up not with the purpose of arguing that studying flora and fauna are not worthwhile topics in a place like Madidi, which indeed they are, but rather to raise the issue of why the impacts of gold mining did not get the same attention from scientists (or, perhaps, from research funders). During a workshop at the Bolivian National Herbarium, one botanist commented that one main reason for this problem is that the vast majority of research projects are determined solely on the interest of the researcher, and the concerns of other local decision makers and community groups are only thought about after the fact, saying, "We have our study topic determined ahead of time, and then we go to a place and have to attempt to get the local people interested in what we are doing. So we aren't responding to any concerns that they might have, and this creates conflict as the relationship isn't an equal one."[32]

Access Points for Impact

One research project was different from the rest: a participatory monitoring project conducted with Takana and Moseténed hunters and fishers along the Beni River that had been carried out by conservation scientists from the Wildlife Conservation Society's office in La Paz about ten years prior to my visit.[33] The hunters and fishers were largely responsible for collecting the data, which included various details of their harvests, such as species, weight, and the number of hours spent hunting. The stated goal of the project was territorial autonomy through providing information that would support sustainable hunting practices in the Indigenous territory, as the Bolivian government required those applying for land titling demonstrate that they were fully capable of managing their own territories in a manner that was economically viable, environmentally sustainable, and able to protect the traditional livelihoods, cultures, and beliefs of their peoples.[34] The data was also interesting to the biologists as a way of obtaining information about the ecological and economic implications of widespread bushmeat hunting.

Initially, I focused on the outcomes of this project. What was the impact in terms of how people hunted or how the information had been used to inform territorial management? But that line of questioning seemed to come up a bit short. The project had been conducted almost a decade prior, and those who had participated had different perceptions as to whether the research had been used to inform decision-making. For example, one older couple didn't seem to understand the main aim of the research ("territorial autonomy") but talked excitedly about things that they had learned, not only how to measure and weigh, but also about the different populations of animals in the forest—which species there were more of, which there were fewer of, and which were disappearing altogether. They also spoke, unprompted, about scientists whose names they remembered and how these biologists, typically Bolivian women from La Paz, had cared about the people in the communities.

One of these women was María Eugenia Copa Alvaro, one of the first people I met in La Paz and someone who had become a good friend. After graduating from university with a degree in biology, Copa had spent several years working on the project, which required her not only to accompany the hunters on their hunts, but also to live in the communities for months at a time. When I interviewed people in the communities where she had done her research ten years before, they spoke about things they had learned while working with her and remembered how she had cried when the project ended and she had to go back to the city. I later visited some of these communities with her and saw firsthand the joy expressed in people's faces upon seeing their old friend.

Copa told me that such experiences were significant for her as well. They changed the way she felt, not only how she felt about how environmental conservation needed to take local perspectives into account, but about people more generally. Prior to her experiences in working with communities, she considered herself to be someone who preferred the company of animals to that of other humans. But living among people who had a different set of values from those she found common in urban settings helped her see not just herself, but her topic of study, in a different light. Copa said: "It was by living with them I learned and grew, I understood a little of a new perspective on life, of a people whose culture was different from mine. I saw how people who hunted valued the animals and the land where they lived in a way that was different from the city and how I had been taught. I came to partly understand it and I loved it, I valued it."[35]

As I began learning more about the project and the ongoing work between the biologists and the Takana communities, I realized that the good feelings between Copa and the communities were important for a different reason. For the members of the communities with whom she worked, Copa was an "access point" to broader institutions and concepts such as "science" and "conservation." For example, although everyone involved in the work was an individual, they were also representatives of their organizations and their communities, which meant that relationships that emerged out of the project could strengthen (or weaken) existing collaborations and alliances. The Takana leadership council had a standing agreement with the conservation science organization that Copa had worked for, and thus the research activities were part of something larger, something that involved not only the specific communities but beyond—the larger Madidi region, perhaps the entirety of the Amazon basin. The lessons that Copa brought back to her colleagues led to similar projects in other places, and the results were a contribution that would be used, alongside other information and through many conversations, to support the rights of Indigenous peoples in Bolivia to keep and manage their land.

In other words, although the end stage of sharing back results was important for all the reasons mentioned earlier, it wasn't the only place where the research "had an impact." Whereas Copa had returned regularly to the communities with the findings of the collected data, the "impact" of the research would not be limited to her sharing of that information and any potential actions taken based on such information. Rather, the process of impacting had already begun before Copa had arrived in the communities. It began even before the organization's scientists met with the Indigenous leaders to decide if and how they would be doing the study, because those decisions depended on earlier conversations about the important research questions and how the interests of the scientists were shared (or not shared) with those of the hunters. And those early conversations had, in turn, depended on other situations—encounters in which scientists realized that it was important to take the interests and perspectives of the Indigenous communities in the region into account.

Furthermore, impact was happening when Copa had accompanied the hunters in collecting the data, when they shared stories about themselves and their families. It was happening when they sat together at meals and when they walked through the forest to get to the lake, talking about themselves and their respective forms of ecological knowledge throughout the process—how she as compared to the hunters understood changes in the environmental based on what they had experienced in their lifetimes. It was also happening when the hunters expressed that their "more practical" knowledge was equal in value to those of the scientists; as one man put it, "the farmers are the most constant researchers, because they are always in direct contact with the earth, with their crops."[36] Another local hunter commented to me that he was also a *licenciado*, a term used in Bolivia to refer to those with university degrees. "You are the biologist of theory, but I am the biologist of the forest," he said.[37] In other words, Copa's presence and activities provided another way of thinking about the natural world—not a better one, just different. Copa explained it as follows:

> More than the end result of the research, what mattered was the whole process. How the hunters were learning to observe and take notes on what they observed. It wasn't so much that the people were learning about the limits of sustainable hunting, which was something that personally interested me, but rather that they were reflecting on things in a different way. They were talking about the findings with each other. So in the end it wasn't just the numbers and information that mattered, but discussing what was interesting or relevant to their lives about the research.[38]

So yes, impact was about the products and outcomes of research, but it was also about what happened during the process of doing the research. It was about interactions and sharing new ideas between people from very different walks of life; impact was something that emerged from the experience for everyone involved. It was also experienced differently for different individuals depending on their role in the project; the experiences of the hunters who were recording their catches were not the same as those of the Indigenous leaders who helped frame the research questions and aims but who weren't involved in data collection. Furthermore, the project was not perceived as universally positive among all those involved; rather, one's particular role in the research largely influenced their individual perceptions.[39]

Thus, instead of impact proceeding in a linear fashion, with a clear beginning and predetermined end, it was more like a network of ever-spreading ties, with innumerable agents, events, and the relationships between them. It was also becoming increasingly clear that impact travels in multiple directions—including back at oneself, as Copa pointed out when she explained how work in the communities had changed her. This kind of impact is exciting but also scary. If we truly engage in impact practices in science, we're destined to be impacted in turn. Our research questions might change, our methods might change, and even our worldviews might change. In other words, we might change. Framed in this way, impact is powerful. And although it cannot be completely controlled or predicted, impact could be navigated, as a skilled sailor would steer a ship through uncharted waters.

The Nature of the Contact

About a year after my experiences in Bolivia, I was back at my graduate university in England, writing up the findings of my research and attempting to make sense of things. As I had begun to understand impact as something that could happen through encounters with access points during the research process rather than just at the end, I needed to explore theories and ideas that went beyond the linear, deficit-focused approaches that dominated much of the literature in my own field. It was around this time that I came across the work of a twentieth-century psychologist named Gordon Allport.

Allport is best known for his work on personality, but he also developed a theory that is still seen to be one of the best ways to improve relations between groups experiencing conflict. This theory, known as the "contact hypothesis," suggests that if certain conditions are met, direct contact between members of different groups (religious, racial, and so on) can lead to mutual understanding and appreciation, which in turn can reduce prejudice and stereotyping. Specifically, contact needed to be between members of equal standing (so that there were not vastly different power levels), there needed to be a strong level of cooperation (rather than competition) and sharing of goals, and there needed to be support from institutional and social structures. If such conditions were *not* met, rather than improving intergroup relations, contact would create even more conflict. In other words, everything depended on the "nature of the contact."[40]

While reading Allport and the work from scholars who had built on his ideas, it struck me that perhaps the type of scientific research I was observing in Bolivia could be a good example of the contact hypothesis in action.[41] When interactions between scientists and community members were laden with power inequalities and misunderstandings, as in many "parachute" research projects, contact deepened resentment and mistrust. But when there was more of a balancing of responsibilities, shared accountability, and institutional support, such as in the project that Copa had participated in, contact could be transformative for all parties involved.

Learning about theories of contact helped me see the potential "impact" of science conducted in a place like Madidi in a new light. I began to rethink the "gap" between research and practice as a series of spaces of encounter or misencounter in which scientists and the public can interact through shared interests, conflicts, and diverse knowledges.[42] In Spanish, the word *encounter* can be translated to the verb *encontrar*, which means not only to meet, but also to find. Thus an encounter invokes possibilities of new discoveries, just as a misencounter denotes a lost opportunity in seeing with new eyes. This reframing helped me understand the importance of thinking about science communication not just as an "exchange" of knowledge, but rather to envision spaces of knowledge interfacing as being about a "process of relating" between different people, communities, and cultures.[43] Also, theories of contact point to the importance of the different *experiences* people have when they come across research and researchers in terms of shaping their broader perceptions of science, something other scholars were suggesting.[44]

This new way of thinking meant that it was important to look more closely at the multiple spaces where the practice of "impacting" takes place to see what was happening there and what kinds of experiences people were having in those spaces. When I began to look, I began to see more clearly. For example, seeing impact as a "process of relating" was integral to research approaches that had existed for decades in the margins of academia under many different names (such as coproduction and action research) but that were underfunded, underutilized, and rarely taught to students.[45] These approaches argued that for research to be of use for decision-making—whether by policy makers, land managers, patients, community members, or practitioners—these groups needed to be involved in key decisions made during the research process. Such approaches have diverse roots in disciplines such as social psychology, critical pedagogy, and Indigenous methodologies, but they have often shared a common aim of elevating the perspectives of local experts to address important issues in a specific place and context.[46]

However, although the scholarship of such approaches is well established, guidance on how to engage in specific, "impactful" research practices is somewhat disparate, spread across many different fields and traditions. It can be hard for an uninitiated researcher to sort through, let alone obtain funding or mentorship to do such an approach well. Furthermore, as more researchers have been encouraged to do "research with impact," much advice has tended toward the mantra of "more is better." For example, certain approaches to framing the degree or quality of public participation in science can give the impression that there is a hierarchy to engagement, in which the "lowest" (least) level is engaging people one step of the process and the "highest" (best) level is when nonacademic actors steer all decisions made in the research. That may make sense in theory, but in practice, it is extremely costly and time prohibitive, not to mention potentially exhausting for all involved, to require the engagement of various publics at all steps.[47]

Furthermore, when we talk about impact, we're often talking about a bunch of different things: getting people to think differently, changing policy, getting more students interested in science, increasing understanding of the scientific process. But some guidance tends to lump different types of public engagement together rather than distinguishing how engagement at different stages of the research could lead to different types of impact. For example, one common form of public participation in science typically limits involvement of laypeople to the data collection stage, which can be highly effective in terms of increasing scientific appreciation and literacy. But this isn't the same process as involving people in developing research questions or attempting to make policy based on research findings. There are many valid approaches to doing "impactful" science, but it's essential to be clear about which ones we're pursuing and to understand that the types of choices we can make will lead to different impacts in real-world settings.[48]

So, rather than trying to do everything, this more targeted approach can help researchers determine what types of impacts they're most likely to see given different interventions. In other words, rather than grouping all "broader impacts" into one bucket, this approach makes explicit that there are different types of outcomes that emerge, and that's okay. We can't all do everything at once, and a given research project probably can't be impactful in every way. But we can determine in advance what our goals are—to influence policy, to increase trust and transparency in the scientific process, to broaden scientific literacy, to fight science denialism, as some examples—and then focus on what works to achieve those goals.

The rest of this book is thus organized in this fashion to articulate the opportunities for impact that occur at different stages of a research project—from formulating questions to seeking consent to collect the data, going through peer review, and beyond. It is my hope that this approach will provide researchers, students, and science communicators with guidance and examples of how different types of impact can emerge from science, and to be impacted in turn.

PART TWO

The Spaces of Scientific Impact

"Knowledge emerges only through invention and re-invention, through the restless, impatient, continuing, hopeful inquiry human beings pursue in the world, with the world, and with each other."
—Paulo Freire

"It is good to have an end to journey toward; but it is the journey that matters, in the end."
—Ursula K. Le Guin

CHAPTER 4

Asking a Good Question

Dave McLaughlin's corn and soybean farm has been in his wife's family for six generations. It is located in central Pennsylvania, a part of the United States where other corn and soybean farms stretch across the hilly landscape for miles. I first met McLaughlin in the summer of 2021 on a visit to the region to interview farmers about their perceptions of scientific research and researchers. A good friend of mine, Molly Cheatum, had been working with area farmers through her role with a local environmental organization, implementing a tree-planting project along farmland streams.[1] I asked her if I could tag along on some of her field visits to meet farmers, and she and her colleague, Bill Chain, kindly helped me set up interviews. Chain worked closely with farmers, and I was particularly looking forward to speaking with McLaughlin, who I was told had strong opinions about scientists.

McLaughlin was one of several farmers in the region who used a practice called no-till, which refers to the practice of planting crops without plowing the soil. No-till is considered an environmentally friendly farming practice because it helps improve soil health by increasing organic matter and reducing runoff of nutrients that can pollute waterways.[2] But McLaughlin was not initially motivated by environmental causes when he adopted no-till practices. He had first learned it from his father-in-law in the 1990s and expanded the practice because it helped him save time and energy. Three decades after he first began experimenting with no-till he was considered a local leader for conservation practices on Pennsylvania farms, someone other farmers looked up to.

Chain and McLaughlin are good friends, but it didn't start out that way. They told me about the first time they met, which was at a Soil Health Promotion Field Day in 2015.[3] Field days are organized by county conservation districts—farmer groups like the Pennsylvania No-Till Alliance or university-run agricultural extension units—to share new research and information on soil health with farmers. In the past, Chain had attended similar events and remembered being impressed with presentations by university researchers on their lab-based soil science research. During a coffee break at that field day, Chain struck up a conversation with McLaughlin and made the mistake of suggesting that such researchers were at "the forefront of soil science."

McLaughlin was not of the same opinion. Chain told me he remembered how McLaughlin had looked him sternly in the eye and told him that the people at the forefront of soil science were *not* the scientists, but rather the farmers who were willing to risk new management techniques to improve their soils, crops, and production. McLaughlin said that, from his perspective, runoff and erosion, which researchers tended to focus on, weren't the main problem. Instead the biggest problem was a lack of water infiltration into the soil, something that he believed got comparatively little attention from scientists.

Now they laughed at their disagreement, and Chain sheepishly said that the longer he'd worked with McLaughlin and other farmers, the more he realized he had a lot to learn from them. That conversation was the start of a new way of looking at the land for him, he said, and he came to realize that environmental scientists and farmers might have similar goals, even if they measure different things. McLaughlin said that his view had also softened a bit. He used to just think that all researchers were out of touch because they thought they knew better than the farmers who worked and cared for the land. But he'd come to understand that the bigger problem was one of funding. Researchers were forced to chase grant dollars to keep their jobs, he said, but from his perspective, such money was typically allocated to topics of scientific interest but that weren't always relevant to farmers like himself.[4]

Chain asked if McLaughlin felt that way about all scientific research. McLaughlin paused for a moment and then shook his head no. There were a pair of researchers who had been different, he said, whose work he considered to be timely and relevant: John Tooker, an entomologist at Pennsylvania State University, and Maggie Douglas, who was Tooker's graduate student at the time. Tooker and Douglas were an exception to the rule, he said.

When McLaughlin first met Tooker and Douglas, he was having a lot of problems with slugs eating his corn crops, and he wasn't the only one. Farmers all over the region, especially those who were experimenting with no-till or reduced-till farming, were having slug issues. The farmers would talk with one another about possible treatments, call up the county extension offices for help, and go on websites like AgTalk for advice. But no one seemed to have the answers.

At the time, Tooker was new at his post at Penn State, which was 50 percent research and 50 percent extension. Extension programs operate across the world and generally refer to research- or university-based education programs that aim to provide technical assistance, particularly with (but not limited to) agricultural issues.[5] I later spoke to Tooker, and he told me that when you're in an extension position, there's a choice that needs to be made: either you have really good ideas about what will help farmers or you need to respond to questions that are brought to you by them.⁶ He admitted to being low on the former, so he opted for the latter and sent a survey around to county extension offices, trying to get a sense of what pest issues farmers were concerned about. Initially, the responses he got back were a bit too vague to direct him on a specific path. But once the word got out that there was a new entomologist in town who was interested in what farmers had to say, Tooker's phone began to ring. And almost every call was about slugs.

A few of the farmers invited Tooker to breakfast to discuss the issue. When he showed up expecting a small gathering, he found that the entire back room of the restaurant had been reserved. More than thirty farmers filled the space, eager to ask him questions about slugs. There was just one problem. Tooker is an entomologist who studies insects. Slugs are not insects, but rather mollusks (octopi and oysters are also in this category). He told as much to the farmers, who were nonplussed. Tooker was asking them what pests they needed help with, and slugs were it. So Tooker had his charge. As he explained it to me, "They were politely asking for help, and I was naive enough to say, okay, yeah, I'll try to help. And so then I went back to my office and I tried to figure out what was going on."⁷

That's where Douglas came in. Douglas was a master's student when she joined Tooker's lab in 2010, and she had always been interested in doing applied research that made change on the ground, not just research that was only interesting to scientists. One of the first questions she asked was, "What are the farmers calling you about?" When she learned that the answer was slugs, she asked if she could help out.⁸

Although Tooker didn't know much about slugs, one thing he did know as an entomologist was that if there was a pest problem, it was likely because something had happened to the pests' predators. Predators play an essential role in the biological control of insect pests and weeds on agricultural fields, and when they are absent, populations of "problem" pests can skyrocket.⁹ Based on this hunch, Tooker got Douglas set up with lab-based experiments, putting slugs in containers with known predators, such as carabid beetles, and soybean seedlings to see what happened. That was when they made a startling discovery: the beetles were all dying.

As it turned out, for her experiments Douglas had unknowingly planted soybean seed that was coated with a particularly potent insecticide known as a neonicotinoid. Neonicotinoids are the most widely used class of insecticides in the world.[10] Although initially they were thought to be less environmentally harmful than other types of insecticides, research has increasingly found neonicotinoids to be toxic to some pollinator species, and they have been linked with large bee die-offs.[11] For this reason, almost all neonicotinoid insecticides are banned across Europe. However, they continue to be permitted for use in the United States, and in some places, such as Pennsylvania, it can be hard to find seed that is not treated with these chemicals.

As Douglas's seed sprouted and grew, the insecticide traveled into the roots and then the leaves of the seedling and finally into the slug as it ate the plant.[12] While the chemical didn't seem to bother the slugs in the least, it was toxic to its predators, and upward of 60 percent of the beetles were impaired after eating slugs infused with the pesticide. Suspecting that the insecticide was the culprit, Douglas replicated the experiment with untreated seed and found that the beetles did not die. Tooker and Douglas followed the lab experiment with a fully replicated field trial on the land of Lucas Criswell, another farmer who was supportive of their research, where they found results supporting the effect under real conditions at scale.[13] Plots planted with neonicotinoid-treated seed had more slugs, fewer predators, and lower yields compared to plots planted with untreated seed.[14]

Tooker knew they had something important to share with the farmers, and when another field day came up, they had their opportunity. They put together a PowerPoint presentation with videos embedded in the slides. Tooker remembered the moment Douglas played a video of one of her lab experiments, which showed a beetle that was clearly dying from having been poisoned by the pesticide. "I still remember the remarks and the gasps in the room of those farmers saying, 'Oh, wow, holy cow,'" said Tooker. "It was clear evidence that something was messing up these beetles. And once they saw that convincing video, the farmers just took it on faith that our data were related to that phenomenon . . . that the reason we were finding fewer beetles in fields was because of that."[15]

For years, there had been significant efforts to get farmers to use fewer pesticides due to their impacts on pollinators, but doing so was a tough sell. One reason is that corn is wind pollinated and soybeans are largely self-pollinating, so whether pollinators were being harmed was not among top concerns for corn and soybean farmers. Douglas's findings began to change that. Suddenly, planting neonicotinoid-treated seeds was not just bad for bees, it was also potentially bad for corn and soybean crops. The research also pointed to a clear solution: seed that was not coated with a neonicotinoid insecticide could be a good option for farmers with slug problems.[16]

However, even as the farmers began to understand that neonicotinoids were part of the problem, a challenge remained: hardly any untreated seed was to be found. Neonicotinoid-coated seeds dominated the corn and soybean markets, and farmers often didn't have a choice over which seeds they could buy for their lands.[17]

This lack of choice got the attention of some policy makers, Tooker later told me. After publishing their work in a top journal in the field, Tooker and Douglas were invited to Washington, DC, to meet with staffers of both the Senate and House Agriculture Committees.[18] Tooker and Douglas prepared a one-pager and began talking about the problem of neonicotinoids, but they could tell from the onset that the staffers weren't very interested in the issue. Neonicotinoids were not a new topic, and the staffers' eyes began to glaze over as the researchers attempted to explain the implications of their findings for no-till farmers like McLaughlin and to put forward solutions. It was clear that they weren't getting their message across. But then Douglas mentioned something about how the farmers didn't have a choice in seed selection, and one of the staffers sat up straight in his chair. "They don't have a choice?!" he repeated. "That's a problem."[19]

Tooker said that the experience was both affirming and disheartening at the same time, because it made clear to the researchers that their science was secondary to the lack of choice of farmers to choose which seeds to plant on their farms.[20] But it also pointed to a potential opening for change, something with which years of lobbying from environmental groups to reduce pesticides used by farmers had not had much success. Douglas told me that in 2021, the Natural Resources Conservation Service started offering farmers a small incentive to plant uncoated seed, citing her and Tooker's research as a key motivator for farmers.[21] She said that in her estimation, one reason their work ended up having impacts for both policy and practice was because it was responsive to the immediate concerns of the farmers with whom they were working:

> When you start where farmers are, you can have a different kind of conversation. Starting with neonicotinoids and focusing on their effects on pollinators does not leave a lot of space for the concerns of corn and soybean growers. Slugs, on the other hand, immediately connect to something they care about. And the no-till farmers we worked with were deeply committed to soil health. So, when we discovered that neonicotinoids were harming soil life and disrupting slug control, this encouraged them to reevaluate their growing practices. It's a discovery that we probably would not have made without the farmers pointing us in that direction.[22]

For Tooker, the research was not just practically relevant, but at its core was an ecological story he also enjoyed sharing with his entomology and ecology colleagues at conferences. One of the research group's papers, published in the *Journal of Applied Ecology*, was designated with an "Editor's Choice," an accolade that distinguishes papers in the journal of especially high quality and interest.[23] Furthermore, the implications went far beyond no-till corn and soybean farmers in Pennsylvania. In 2018, a paper came out in Australia that found similar slug-beetle-insecticide interactions from spray applications of neonicotinoids.[24]

When I asked Tooker why he thought the farmers accepted his recommendations, he said the farmers had never asked for the insecticides to be put on their seeds, so Tooker and Douglas weren't directly critiquing any practices the farmers had intentionally chosen. Perhaps more importantly, however, Tooker said he believed they accepted the results because the farmers had been the ones to bring the question to him. He speculated what might have happened if, doing the same research and using the same approach, he and Douglas had come up with the question, and he guessed that it might not have been so well received:

> If I had come up with those ideas on my own and followed a similar research path, maybe not as engaged with the farmers, I don't think it would've had the same results, the same outcome, or the same impact, because my motivations could have been questioned. Maybe I was anti-insecticide, maybe I was a tree-hugging environmentalist. Maybe I was just looking for an interesting story. You could come up with a variety of reasons for why they wouldn't trust me. But why they did had a lot to do with the request and our willingness to act on that request.[25]

As for McLaughlin, Tooker and Douglas's findings changed his view—albeit slightly—about the potential of researchers to do something that helped farmers. McLaughlin told me that he trusts Tooker and Douglas's expertise. If they make a suggestion, he considers it worth trying, he said.[26] And to this day they are friends. When Douglas got a faculty position at Dickinson College, McLaughlin invited her to bring her students to his farm to teach them about the importance of focusing on soil health for combatting erosion and runoff, visits that continue to the present day.

Eight years later, neonicotinoid-coated seeds are still standard for corn and most soybean, but that's slowly changing as more noncoated seeds are made available for farmers to purchase.[27]

Tooker and Douglas's work shows both the importance of involving those who could potentially use the research in the development of research questions and the limitations of such approaches to make large-scale change in the face of systemic challenges. However, states that border Pennsylvania, such as New York, have passed bills to prohibit or limit the use of neonicotinoids, drastically changing the planting landscape for farmers.[28] Tooker and Douglas's work has been part of those conversations, which at some point may begin happening in Pennsylvania as well.

In the meantime, farmers like McLaughlin are moving forward with evidence-based techniques. McLaughlin now uses only untreated seed for his soybean and wheat crops, a switch he made in large part due to Tooker and Douglas's research. This change is important not just for McLaughlin's farm but for others in the region, as one of their most important information sources is other farmers—particularly neighboring farmers. Chain told me later that this information flow is referred to as the "twenty-mile rule": if a neighbor within twenty miles is doing something and it is working, that is a compelling reason to try it on one's own farm.[29] Indeed, McLaughlin said that he was part of a "community of peers" who supported one another in trying out new techniques and that the number of farmers in the region opting for integrated pest management approaches, as opposed to conventional insecticides, is growing.[30]

Who Asks the Questions?

The story above reveals that one of the most important ways that research—and researchers—can have impact is to those who could potentially use the findings of research, such as farmers or practitioners, about what is important to them and then "take on the charge" by focusing one's research on that topic. Even though this approach may seem like common sense, especially to those outside the academic community, it is relatively rare for several reasons.

One reason is how we are trained to do research. Any student of science knows that one of the first steps of the scientific method is to ask a research question. But how does this step typically play out in science? How do we pick a research question?

When we are learning about science and research, we are often steered in the direction of what is easily measurable. In middle and high school, most often we ask questions that have already been answered for the mere experience of completing the experiments that have led to our most basic scientific foundations. We learn about the physics of gravity, for example, by dropping two objects of differing mass from a high point, or we simulate the greenhouse effect by using baking soda and vinegar to trap carbon dioxide within a jar. For teaching purposes, this approach is useful because, as educators, we already know the outcome and can anticipate the questions students might have.

As we proceed in higher education, we often work under the guidance of academic mentors, who often provide our research questions for us, which makes a lot of sense. Research is extremely labor- and resource-intensive, and developing a good research question requires not only deep knowledge on a particular topic, but an understanding of what answering that question would entail. Having taught multiple iterations of research-based courses to both undergraduate and graduate students, I get the practicality of this approach. Most students struggle greatly with coming up with an answerable research question on their own. They will often pick questions that are too vague to be answerable or that we do not have the time or resources to address. Thus, when mentoring students, it is often much easier to point them in the direction of something that is small, measurable, and falls within the realm of our own disciplines than let them fumble about on their own.

But there is a trade-off to this approach: through this process, students are often led to assume that the best research questions come from within academia. As a result, important ideas, some that are based on diverse life experiences, will inevitably be missed. As discussed in chapter 1, each of us, no matter how brilliant, is limited in how we see the world. Our prior education and experiences train us to see some questions as important and not others, and questions that seem "less academic" or that aren't perceived to contribute to the scholarship within our own disciplines are sometimes discarded as not being worth our attention.

Indigenous scientist and botanist Robin Wall Kimmerer describes the impact of this limited approach in heartbreaking detail in her beautiful book, *Braiding Sweetgrass: Indigenous Wisdom, Scientific Knowledge, and the Teaching of Plants*. As an entering freshman in college, she had chosen to major in botany because she was fascinated with plants, and she wanted to know the answer to a research question that occupied her thoughts: why do purple asters and yellow goldenrod look so beautiful together? But when she shared this question with her academic advisor, her enthusiasm was met with disdain:

> Why do they stand beside each other when they could grow alone? Why this particular pair? There are plenty of pinks and whites and blues dotting the fields, so is it only happenstance that the magnificence of purple and gold end up side by side? . . . But my adviser said, "It's not science," not what botany was about. I wanted to know why certain stems bent easily for baskets and some would break, why the biggest berries grew in the shade and why they made us medicines, which plants are edible, why those little pink orchids only grow under pines. "Not science," he said, and he ought to know, sitting in his laboratory, a learned professor of botany.[31]

What Kimmerer's passage points to, among other lessons, is the tendency for many of us in academia to assume that the best research questions, the most valid questions, are those that emerge from prior scientific research, particularly research in our field. After all, we are the ones steeped in not only the foundational research, but in the cutting-edge theories and methods in our disciplines. We will come up with the most innovative knowledge and the best solutions to the world's problems—as long as society leaves us alone to do our thing. In other words, we know best.

This vision of science became prominent at the end of World War II, most notably in a report written by Vannevar Bush, science advisor to then US President Harry Truman. Bush argued that "scientific progress on a broad front results from the free play of free intellects, working on subjects of their own choice, in the manner dictated by their curiosity for exploration of the unknown."[32] Bush's report was hugely influential, directing much investment for research conducted at colleges and universities toward "pure" science as compared to "applied" research.[33] While "pure" science is hugely important in that it can lead to novel discoveries that underpin new technologies and medicines (for example, see the Golden Goose Awards[34]), the time frame of impacts that emerge from basic science can be very slow.

In his article "Saving Science," science policy researcher Daniel Sarewitz argues that although scientists are more productive than ever, publishing millions of peer reviewed articles each year, the pursuit of scientific knowledge has become increasingly disconnected from addressing real-world problems and is thus often unusable.[35] Rather than funding "the free play of free intellects," Sarewitz argues that some of the greatest scientific and technological advances have emerged from research structures that have involved collaborations between scientists and interested groups from various fields to solve complex, real-world issues. For example, he tells the story of how the participation of breast cancer patient-activists in a grant merit review process led to funded research on a new therapy and the creation of the drug Herceptin, currently one of the most important methods of treating breast cancer.[36]

Thus, rather than protecting science from societal influences, "science will be made more reliable and more valuable for society today . . . by being brought, carefully and appropriately, into a direct, open, and intimate relationship with those influences."[37]

The Power and Peril of the Group

Although Sarewitz doesn't directly say it, his argument is largely based on the importance of dialogue and differing perspectives for scientific progress. Scientists have long understood the power of communication and collaboration with other scientists as a crucial driver of innovation and new ideas. Long before the internet, researchers communicated with one another via letters, and the ubiquity of scientific journals and meetings is one of the pillars of scientific progress.[38] A new theory or discovery gains value when it is shared with others who can learn from it and who then build upon that knowledge in new directions.

Communication and collaboration are thus two hallmarks of scientific research, not only important for getting ideas out, but for bringing new ideas in. One way science advances is when unexpected findings challenge prior assumptions, providing the clues that the previous way of thinking is outdated and needs to be questioned. As famously argued by science philosopher Thomas Kuhn, when enough anomalies within a prevailing scientific paradigm accumulate, a new paradigm emerges, fundamentally altering the way scientists perceive and interpret the world.[39] In this process, lone geniuses with brilliant ideas are less important than dialogue and debate between diverse communities of knowledge—for example, when one way of thinking encounters another approach.

Interestingly, this progress mimics human evolution, which largely advanced because of collaboration and what happens when communities of brains work together.[40] Cognitive scientists have often found that people are "smarter" when they work in groups as compared to on their own; they more consistently find the answers to logic-based problems and are less likely to resort to superficial explanations.[41] For example, one of the most well known tests of deductive reasoning, the Wason selection task, presents a participant with four cards, each with a number on one side and a color on the other. The participant is told that the cards should follow the rule that if a card shows an even number on one face, then its opposite face should be blue. Which card, or cards, must the participant turn over to know if the rule is true or false? (If you're curious about how this works, Google the task and try if for yourself before reading on.)

Researchers have consistently found that individual participants perform poorly at this task, especially when it is presented without any context for what the numbers or letters mean, because the answer requires going against intuitive reasoning; to solve it correctly, it is necessary to think through the problem carefully. In Wason's original study in 1966, which was replicated with similar results several decades later, only about 10 percent of participants could correctly solve the puzzle.[42] But research has found that one reliable way to increase performance is to have people solve the task in groups rather than alone. In one iteration of Wason task, the percentage of participants who were able to correctly solve the puzzle increased from 9 percent to 70 percent (!) when people worked in groups rather than individually.[43]

However, there's a flip side to the power of collective intelligence. Groups that are made up of only like-minded individuals (for example, echo chambers) can reduce the gains in reasoning that are observed in collective settings, leading to groupthink, where arguments will not be critically evaluated.[44] Groupthink tends to preserve the status quo rather than exploring novel ideas, thus restricting innovation and progress.[45]

Similar to other groups, scientists who share ideological or field-specific worldviews are susceptible to gatekeeping and groupthink.[46] Scientific theory can quickly become dogma without regular debate, critique, and argumentation.[47] For example, in recent years, there has been a reckoning within the medical community working on Alzheimer's research. For decades, one theory of Alzheimer's disease (known as the amyloid hypothesis, which theorized that the accumulation of amyloid plaques in the brain was the primary cause of Alzheimer's) dominated the biomedical community. Proponents of the hypothesis largely determined which research projects got funding, whose careers were advanced, and which drugs were supported for clinical trial while critics of the amyloid theory saw their research go largely unfunded.[48] Only in recent years has there been a growing acknowledgment of the need to explore diverse research avenues beyond the amyloid-centric approach.[49]

One essential way to circumvent groupthink is through diversity in the broadest sense. Much research has shown that human diversity in terms of gender, race, and socioeconomic background, among other factors, can provide gains in collective decision-making.[50] Scientific teams with greater representation of ethnic diversity have been found to be more innovative and to publish articles that receive more citations as compared to less diverse teams.[51] In scientific communities, diversity can also refer to epistemic diversity (differences in perspectives and knowledge), which can be supported through talking with researchers from other disciplines.[52] But although calls for interdisciplinary research teams are common, what is less discussed is the value of incorporating perspectives from outside the academy into the design of research projects and the selection of questions.[53]

Cognitive scientists would argue that this omission is a lost opportunity. Interactions between researchers of different disciplines and backgrounds can help to break through groupthink in terms of scientific theories, but there still exists a limitation in that all those involved are addressing the issue from a shared academic lens. Involving laypeople, particularly those with a stake in the matter, can, however, shine a light not just on what information is needed to address a problem, but how or whether a proposed solution can be implemented in practice.

Questions, and Impact, Are Context Specific

Matt Bomgardner is the owner of Blue Mountain View Farm, a dairy farm that recently made the transition to organic production. Switching from conventional to organic is no small move and requires huge investments not just of time and money, but also energy, curiosity, and even the ability to withstand social pressure. The farm was among those included in the tree-planting project led by Molly Cheatum and Bill Chain, and I accompanied my friends there on a hot day, the mercury pushing ninety degrees Fahrenheit on the thermostat as their team measured the growth of the trees planted the prior year. Bomgardner wanted to observe their progress, so we stood in the thin shade of trees along the creek that cut through his land. He told me that he had selected the spot where the team was working because it was in an area that was not one of his prime pastures. It was important to see how it would work in a less valuable part of his land first, he said; if he was happy with the results, he would plant more trees elsewhere.[54]

Bomgardner told me that determining what was best for one's land was challenging because everything is context dependent. Other farmers are one important source of knowledge, but sometimes they don't know things either—in general, it is best to see for oneself what works and what does not. For example, he mentioned a certain book about grazing techniques and said that although its practices were appropriate for temperate climates such as New Zealand and Ireland, they didn't work for Pennsylvania, which has hot summers. He explained that in some systems, it's important to have short grass, whereas in others (such as his farm), taller grass is better because the cows stamp it down into the soil. Everything needed to be tested—it wasn't smart to just take someone's word for it, he said.[55]

"Not taking someone else's word for it" reminded me of a phrase I'd learned from research into the history of science. *Nullius in verba,* which roughly translates to "on the word of no one," is the motto of the Royal Society, which was established in London in 1660 and is generally considered to be the oldest scientific society in the world.[56] Not just a place for the sharing of ideas, the Royal Society was set up to put scientific ideas on trial, quite literally, as "men-of-learning" performed experiments in the private rooms of wealthy aristocrats or in small college laboratories.[57] The society's motto underscores the necessity of a system in which scientific discoveries could be shown and verified by one's peers, thus allowing others to see with their own eyes whether something could be shown to be true rather than taking one's word for it. So Bomgardner and other farmers who also wanted to see if something worked before believing were in good company.

Another farmer who emphasized the importance of testing ideas for oneself was Leslie Bowman, one of thirty-four family members working a large egg and grain farm that has been in his family since the 1800s. We pulled up to the farm to find Bowman standing at the bed of his pickup truck, looking carefully at four small bundles of newly cut corn. We shook hands, and Chain pointed to the corn and asked him what it was for. It was an experiment, Bowman explained. He had cut the corn earlier that day to determine the precise concentrations of trace minerals, such manganese, copper, and zinc, in his plants. We followed him into his office, where there was evidence of past experiments—dried soybean plants on his desk, plastic containers with dried ears of corn. Everything had a purpose, I soon learned. Bowman picked up the bean plants to show us the nodules on the roots, which he explained were able to remove nitrogen from the atmosphere and "fix" it in the soil.

As I held up the bean plants close to my face, squinting to try to make out the nodules, he sat down at his desk, where a massive computer screen projected a high-definition map of his entire farm. The map was built using multiple GIS layers and provided a meter-by-meter analysis of soil productivity of his entire farm. Using this detailed information,

Bowman explained that he could make decisions about how many seeds to drop in each area to make sure he wasn't putting too many seeds in places where there wouldn't be enough resources to ensure that each would have enough to grow well. The maps also demonstrated the past use of the land—for example, where a road had once passed through and thus where the topsoil needed to be reestablished.

As an outsider and city dweller, when I looked at the farm, I saw corn and beans. Bowman, though, saw intricate relationships between soil, plants, and air happening on a scale best studied by the square meter. That was quite a feat, considering that his farm was 2,370 acres in size. And although none of the other farms we visited that week had drone photography technology, all knew their land with the same type of intimacy. They knew which parts of the land had compacted soil, which were prone to invasive weeds, and which would be waterlogged after a heavy rainstorm. In other words, the farmers had deep knowledge that was extremely valuable for knowing how things would work (or not work) on their particular patch of land.

Increasingly, there is awareness in academia about the importance of other types of knowledge, such as that held by farmers or other land stewards, including Indigenous communities. Such knowledge is valuable in its own right; as pointed out above, it can improve the precision of a broader understanding of a system or even demonstrate when scientists have misinterpreted interactions on the land.[58] However, there are still major challenges in terms of how to "integrate" different knowledge systems in the practice of research.[59] One area that has been particularly problematic, as pointed out by McLaughlin during our initial conversation, is funding. Research funding typically privileges research projects that are geared toward the creation of broadly relevant findings rather than toward what works in a particular region. But there is a mismatch here in terms of extra academic impact because the potential of those findings to influence specific practice and policies is almost always a matter of how things work in a very particular place and time. For example, it might be useful to know that a novel method of sowing might improve yield, but as the farmers I spoke with pointed out that to implement that particular innovation, a farmer would need to see exactly how it could work on their own land.

Some funding approaches do have the potential to lead to actionable research. For example, Tooker was able to spend time engaging farmers because his extension position encouraged him to do so; extension programs provide the infrastructure and incentive for supporting researcher-farmer relationships. Furthermore, the initial trials conducted by Douglas were funded through a special grant program designed to support farmer-driven research. The program, called Sustainable Agricultural Research and Education (SARE), is funded by the United States Department of Agriculture's National Institute for Food and Agriculture and includes farmers, ranchers, and others who work in the agricultural sector in their grant review panels, which means that it is not only researchers who evaluate the potential value of the research.[60] Furthermore, larger funded projects are required to have advisory boards consisting of diverse members of the agricultural community to oversee and guide their direction.[61]

The involvement of practitioners and other interested "users" of research in the funding process also helps circumvent the problem of groupthink described earlier because it inserts diversity of expertise and background into the equation. That's an essential point; in the past, homogeneity in who decides what should be funded has led to missed opportunity in terms of topics, especially in health and environmental sciences. For example, for decades in the medical community, there has been a documented lack of research on major women's health issues, such as menopause.[62] Gender bias in the grant review process, which has historically been male dominated, has been cited as a factor in the disproportionate share of funding for diseases that affect primarily men as compared to diseases that affect primarily women.[63] Similar trends have been found in comparing funding between sickle cell anemia, which primarily affects Black patients, and cystic fibrosis, which is more common among White individuals. Although sickle cell anemia is three times more prevalent than cystic fibrosis, it has received far less funding from both public and private entities.[64] These examples point to the crucial role of diversity—of race, background, and training, among other factors—in the makeup of grant review panels. As Douglas put it:

> I think it's really important to invest in the extension system and innovative funding models. I love that [SARE's] funding model formally involves farmers very early in the process so that they actually have a way to weigh in and say, what research do we think is valuable? What do we want to see get done? I think that's appropriate. These are public funds, and so the public should have a voice in deciding how those funds are spent, especially the people who the research is hoping to influence or inform.[65]

Asking Questions about Questions

During my time in Pennsylvania, I had the good fortune to speak with almost a dozen farmers, and I was struck at how they were all similar to and yet so different from one another. Each farm was its own country, its own ecosystem. Similarly, each farmer had their own approach, dependent on a multitude of varying factors, including time, amount of land, money, help from family, market connections, external concerns, age, energy, and personality. But what they had in common was that they were constantly observing, constantly asking questions. They wanted to know what would make their crops grow better, their soil healthier, and ways to do things more sustainably (if only to get the environmental regulators off their backs). Tooker told me that doing research with the farmers' interests in mind helped him not only to do work that was relevant for policy, practice, and science, but to teach him more about the value of what they do and who they are as people. And speaking with Douglas, she said, "What I learned from this experience is that, as a scientist, I need to meet people where they are and be open to learning from them, whether I am working with farmers or policy makers. They are responsive to their immediate realities. Understanding that helps to develop work that matters."[66]

We don't know what we don't know, and it's only when we begin to admit and articulate our uncertainty that the path of learning begins. Asking why we ask which questions, and getting our collaborators to ask why they ask their questions, can shed light on the potential for impact at this early stage of the research process. Botanist Mary O'Brien wrote: "It is an interesting exercise to examine the questions you are pursuing as a scientist. Who wants me to be looking at certain kinds of questions, and why? Whose questions am I ignoring? Who is being hurt on Earth, and whom am I trying to save? The muttelet? Mink in the Columbia River? Asthmatic children in inner cities? Mexican workers in border-town factories? The forests? The ozone layer? Farm workers? Groundwater? Biodiversity? No one?"[67]

I asked Tooker what his next big question was. What were farmers calling him about now? He said he didn't have anything big yet in the pipeline. He was getting some calls about different things, but there wasn't a clear issue emerging yet that was shared among the farming community. I asked him if perhaps he could organize a workshop to get farmers together and brainstorm the most important questions. "Nah," he said, explaining that he felt that approach would be artificial; it wouldn't get at the most pressing needs because those emerge organically. He would just be patient, he said. The farmers would let him know when there was something big and important, and he said he would be there, ready to listen to their questions.[68]

CHAPTER 5

The Privilege of Choice: Methods, Permissions, and Location

WHEN I WAS A GRADUATE STUDENT and about to head to Nicaragua to do research for my master's thesis, one of my advisors gave me a piece of advice: "Just remember the first tenet of research: 'do no harm.'" I vaguely remember nodding at him, as if I understood, but also thinking, quite naively, that of course I would do no harm! After all, I wasn't doing any ethically questionable biomedical research or psychological experiments, which research ethics committees were first established to oversee.[1] Rather, I was just going to do interviews with people and perhaps a few, nonintrusive ecological measurements. What could be harmful about that?

I was twenty-five years old at the time. I remember my excitement in traveling to a new place, where the people had lives so different from my own. I hoped to learn something new and share that knowledge with my friends, family, and colleagues back home. Close to twenty years later, I now see what I was: a "parachute researcher" following in the footsteps of countless others from the Global North to make "discoveries" in the Global South, without much prior thought of how I could do research that equitably engaged with the existing knowledge of people in my "field site."[2] I was ignorant of the potential negative impacts that my researcher presence could cause, believing as I did at the time in the idea that the practice of scientific research is one that is "neutral." This idea, which I likely picked up from the way research was talked about in my university courses, has been passed down from generation to generation of researchers. Linguist Mary Louise Pratt pointed this out in her recordings of early naturalists' writings about their scientific endeavors in Africa and South America: "[In Africa], the naturalist will find a vast field for his observations, and there he will discover objects capable, by their immense variety, of satisfying all his tastes.... Penetrated by such sentiments, and greatly excited by the perspective of a land whose products are unknown to us, I left England with the resolution to satisfy a curiosity which, if it is not seen as useful to society, is at least innocent."[3]

But as many have argued, scientific research is never neutral or innocent.[4] For example, science brought us eugenics, which served to justify the use of harmful methods such as involuntary sterilization as a means of ridding society of the "unfit."[5] Science has perpetuated social inequalities, providing now-debunked "evidence" to justify racist beliefs about the intellectual inferiority of racial minorities, greatly devaluing the contributions of people of color in science.[6] Scientific research has also often been a practice of extraction—natural history museums around the world are full of the fruits of scientific labors, including cultural artifacts, biological specimens, and even human bones, taken from one country to be displayed in another.[7] Even when researchers stay within conventional ethical bounds, parachute science can hinder locally grown research efforts by creating dependency on external expertise and prioritizing research topics that may be interesting to foreign scientists, but that do not address local research needs.[8]

These legacies are not relegated to the past but, rather, have implications for the practice of research in the present day. For example, who gets to do research is largely a matter of access to educational opportunities and positive experiences in academic settings. Disparities in educational opportunity are one reason, for example, that scientists around the world are often more likely to be from socioeconomically privileged backgrounds, particularly the higher up in rank one goes.[9] Another consequence of such legacies is that the research institutions with the most resources are concentrated in high-income countries; researchers from these countries are also disproportionately listed as first authors in the top journals, even when writing about research conducted in the Global South.[10] As such, the parachute researching I was funded to do in my midtwenties is still common, despite many calls to end the practice.[11]

These disparities mean that the practice of doing science comes with heavy baggage and related responsibilities. But it does not mean that there are not ways to do science that can lead to more equitable outcomes. Scientific research, regardless of discipline, is above all a series of choices. Some of these choices are directly within our power (such as whether to do research at all), and some are not (which methods are appropriate for a given field of study). Some choices are largely individual, and others are driven by funding availability, institutional infrastructure, and the norms and ethics of scientific societies. These choices alter from place to place, among disciplines and levels of seniority, and change over time, so the choices faced by, say, an Egyptian graduate student of botany in the 1980s will not be the same as those faced by a tenured sociologist from Australia in 2024. In other words, the choices that one researcher faces will not be the same as those faced by another. Regardless of location, discipline, personal identity, or time period, however, in research we must make many choices, and what we decide has important consequences in terms of the types of impacts that our research may have.

Over- and Under-Researched Places

One of the most important choices we make in research is location. All research is conducted in a particular place, whether it is a fluorescent-lit laboratory in the basement of a university science building or atop the melting glaciers of Antarctica. Typically we choose the location of our research based on a few different criteria. One factor is usually convenience, especially if we're limited on budget. We typically do research where we have access and where we can afford to spend time. If we do lab or computer-based research, location is typically a matter of where our institution is based and the available facilities. A second key factor is methodological rigor. If fieldwork is required for our research, we often need to choose a location based on the presence or absence of certain variables. For example, if doing a natural experiment in social science, we might need to find two groups of people with different circumstances, perhaps because of a divergence in law, policy, or practice between cities or neighborhoods. Similarly, ecological research often requires the selection of a location with specific characteristics related to our topic of study.

Once these two factors are accounted for, there is still often choice, and so an additional factor that typically determines where we go is based on interest: we go where we are interested in going. I think we need to think a bit more critically about this factor. For example, the prevalence of parachute science is largely due to the very unequal power relations inherent in where we choose to do our research. With this choice comes a certain type of privilege—an ability to buy plane tickets and obtain visas, for example—that is largely based on power derived through one's nationality. This power imbalance recalls a scene from a play written by Robert Ajwang', a dancer and musician from Tanzania, and American theater scholar Laura Edmondson—a husband-and-wife team who met when Edmondson was doing research on Tanzanian popular theater and

Ajwang' worked as her research assistant. In the scene, Edmondson explains her choice in picking Tanzania for her field site:

> I am frequently asked, Why Tanzania? Why not Nigeria, Ghana, Senegal, Cote d'Ivoire, Uganda, Botswana, the Central African Republic? In response, I usually speak of its strong tradition of Swahili theatre, which has been frequently overlooked by those Eurocentric researchers whom I so carefully distance myself from. I was to be adventurous. A pioneer. Going where no ethnographer of African theatre had gone before—
> (Robert snorts and starts counting on his fingers.)
> ROBERT: Well, let's see, we had Jane, Matthew, Claire, Tim, Mark, Kathleen, George—[12]

Ajwang' and Edmondson's play also points to the notion of some locations being recipients of too much research of the same type. The term *over-research* has been used anecdotally in the research community for a long time, but, ironically, it is an understudied phenomenon.[13] Over-research is often used to describe locations, communities, or topics that have received a disproportionate share of attention from researchers, while other places and issues are relatively neglected. This can have negative consequences for science, as the overabundance of research in some places can represent a type of research bias, thus potentially skewing data and leading to misinterpretations about people and places.

Over-research has been raised as an ethical concern in research, and it can have major impacts for communities.[14] As I found during my PhD work in Bolivia, communities that experience a steady trickle of researchers—many of whom may only stay for months or weeks and who do not return the results of their findings—can lead to resentment, suspicion, and, ultimately, closed doors. Even so, over-research remains absent from most ethics and funding guidelines. As Cat Button and Gerald Taylor Aiken articulate in the introduction to their book collection on over-researched places, "Over-research seems to be something reflected on when pointed out, but not an issue considered in the planning stages of many projects."[15] They also find that locations can be over-researched in terms of topic, methods, or approach, but neglected in other respects.

This point brings up the counterpoint of some places and topics being "under-researched" due to remoteness or lack of attractiveness or because the people, culture, or environment are perceived to be less "interesting" than other places. For example, urban geographer Hanna Ruszczyk wrote of the many newly formed midsized cities across the Global South that often fail to attract researchers because of their "ordinariness." Of such cities, she said that historically, they "have not been interesting to research. They are harder to reach (both physically and mentally) and they do not have the high voltage power of the capital. Ordinary, boring cities that have no particular claim to fame do not have researchers clamouring to research them."[16] That's a mistake, she argues, because such cities may be more representative of how people live compared to the megacities where the majority of researchers and research institutions are located. She wrote that "to overlook is not merely to ignore. It may involve a conscious choice to look elsewhere or it may constitute an act of simultaneously knowing but not caring."[17]

A similar argument can be made for the field of ecology, which for decades neglected urban and suburban environments in favor of rural, "pristine" landscapes because the former were frequently considered to be devoid of nature.[18] Thankfully, this omission has largely been rectified within the ecological community, where studying ecology-society interactions in urban settings is increasingly seen to be important as the majority of people now live (and have nature-based experiences) in cities. The expansion of ecological research into human-dominated environments has contributed greatly to new theories, interesting questions, and the overall growth of the field.[19] However, recent systematic reviews have found that location biases persist in other areas of environmental research, such as research on climate-driven conflicts, where scholarly attention has focused more on rural as compared to urban dynamics.[20]

Over-research and under-research are also seen in other areas of science, such as clinical trials, where people of color are routinely underrepresented in opportunities to participate in medical studies.[21] As a result, medical information is not always meaningful or useful to "all" people. For example, darker skin tones are underrepresented in medical textbooks, which can lead to misdiagnoses and bias in treatment for patients.[22] A recent study found that Lyme disease shows up differently in individuals with darker skin, but images of ticks on black or brown skin are virtually absent from medical databases, let alone available on the internet.[23] Technologies are similarly often designed for use by majority cultures, neglecting potential differences in how they would be used by members of minority groups. For example, pulse oximeters, which are commonly used to measure a person's blood oxygen level, have been found to be less accurate in people with darker skin, leading to delayed medical treatment for people of color as compared to White individuals during COVID-19.[24]

Building Equitable Collaborations

Funding organizations have a large role to play in whether certain topics or locations are researched and can help to reduce existing disparities. For example, recently there has been an important call to invest in science done in Africa by African scientists (at African institutions).[25]

Such initiatives can retain talented and committed scientists in their home countries and drive technological innovation and social progress. This practice is fundamentally different than funding scientists from the Global North to travel to Africa to do their research. However, it may be challenging to implement, as some scientific funding bodies limit financial support to researchers based in their home country.

Another consideration is that of cultural competency, which refers to one's ability to navigate the language, norms, and rules (many of which may be unspoken) of different cultures.[26] Such competency is particularly important in places where there may be a large power differential between the arriving scientists and the people who live or work on the land. For example, many Indigenous communities around the world have played host (often unwillingly) to scientists from diverse natural and social science traditions.[27] However, Indigenous people are underrepresented in terms of having formal positions in research.[28] Doing research on Indigenous lands or with Indigenous peoples without a deep understanding and respect for the unique histories, knowledges, and cultures of such communities can cause unintended harm. For example, in 2003, members of the Havasupai tribe in Arizona learned that their genetic samples, which they had given consent for researchers from Arizona State University to collect for diabetes research, had been used for other studies, including mental health and inbreeding studies, without their approval, leading leaders of the Havasupai to issue a "banishment order" and refusing any Arizona State employee entry to their lands.[29]

In such contexts, cultural competency is an essential ingredient for building trusting and equitable research relationships. Understanding of language, norms, and rules is needed to obtain true informed consent, which is an ethical imperative for the research community. For example, for many Indigenous communities, biological samples, such as blood, are seen as sacred. In this sense, researchers who seek consent for such samples must understand their role as respectful stewards with obligations to those who gave their samples.[30]

Despite the importance of cultural competency, it is not currently a requirement for most research funding. For example, I obtained prestigious funding awards both for my master's research in Nicaragua and my PhD research in Bolivia without ever having stepped foot in either country. Strikingly, despite doing research that involved local communities, nowhere on either funding application did I even have to indicate that I had awareness or understanding of local cultures and languages.[31] Indeed, over the years, I have come across foreign social scientists in various parts of Latin America who have only a rudimentary grasp of Spanish. Going into such situations without proper training and awareness is akin to sending a student into a laboratory without protective equipment, safety protocols, and instructions on how to dispose of biohazardous waste. In other words, it would not be acceptable in even the most relaxed lab settings. So why is its cultural equivalent still the default across much of the research community?

One way to increase cultural competency in science is to make sure that research teams include members of communities where research is being conducted. Much research has found that hiring locally is not just good practice, it's good science. Local field assistants can provide important local knowledge, but access to such knowledge should also be fairly compensated. Anthropologist Hugh Raffles noted this point in his book *In Amazonia* in which he describes how one guide became so skilled that scientists no longer wished to hire him. He wrote: "From the first day the irony begins to unfold. The more successfully [the guide] teaches, the faster the visiting scientist learns to be independent. Within a couple of years, the project is established, the team is trained, the area mapped, the replicates in place, and everything has settled into a secure routine. Once indispensable, his skills are no longer needed. And his price, which has risen steadily over the years, is too high."[32]

As Raffles pointed out, determining fair wages in research can be incredibly tricky and context dependent.[33] Researchers work in all types of environments and across national boundaries. For example, a researcher from, say, Belgium, does not have to (and perhaps should not) use minimum wage scales based in Belgium for compensating day labor of research assistants in, say, Guyana. On the other hand, the researcher shouldn't only use local labor laws, which may vastly undervalue the work being done, as the main metric for compensation.

Another potential issue to consider is when researchers wish to pay community members with limited resources to assist in the research process (as community-based researchers). It can be a great opportunity, but it doesn't come without pitfalls. For example, many people with limited income rely on governmental assistance in the form of medical benefits or financial aid. Such individuals may risk receiving less in their monthly assistance check, or losing benefits altogether, if they exceed their earning exemption limits.[34]

Some research centers and institutions have created guidelines for paying research assistants, particularly those who may come from under resourced communities. For example, the University of British Columbia Centre for Disease Control has an excellent guide for best practices in paying community-based research assistants located in the city of Vancouver. The research includes guidance for research leaders in terms of thinking about barriers to equitable pay, such as lack of access to bank accounts (and examples of low-barrier banks in the region), issues with obtaining identification for undocumented individuals, and how to provide a letter of support for a prospective assistant to have income exempted.[35]

Permissions and Consent

A related aspect of choice deals with permissions and consent. The subject of research permissions—whether to obtain ethical approval for conducting human subjects research, to obtain biological samples, or to move organic material from one place to another—is often viewed through the lens of bureaucratic box-ticking, something to get out of the way to be able to get the work done. Indeed, on many grant applications, the question of whether a researcher has gotten a specific type of permission to conduct the research is an actual box that must be checked before data collection can begin. And perhaps because it looks like a bureaucratic exercise, it often feels like one in practice.

That's unfortunate, because it can lead us to forget why we are seeking ethical approval in the first place, which is to ensure that the risks of research do not outweigh its benefits. It's important to remember the past injustices and atrocities committed in the name of scientific research, such as the well-known Tuskegee experiment than condemned hundreds of African American men, and many of their family members, to live and die with untreated syphilis.[36] As discussed in chapter 3, scientific "impact" is not always positive; research is not inherently good for all people, and thus neither its benefits nor its risks will automatically be shared equally across society. Because of this potential for harm, scientists must be reflective of the potential risks that their research may carry and act accordingly. That's challenging to do, as risks that emerge from the production of knowledge cannot always be predicted, just as Albert Einstein never could have anticipated that his theory of special relativity would have ultimately been used to kill a third of the population of two Japanese cities. But in the 1980s, some scholars began thinking more deliberately about risk emerging from science and technology and developed theories of how it might be best managed.

One well-known approach was developed by sociologist Ulrich Beck, author of the book *Risk Society*, one of the most cited academic books of all time. Beck argued that although contemporary society has largely benefited from technological improvements, scientific discoveries, and medical advancements, such benefits have also come with a high degree of risk and uncertainty.[37] Because such risks are not "natural" (like an earthquake), but rather human-caused, it is up to the societies that create them to determine when potential benefits outweigh risks, and vice versa. Beck, and other scholars, argued that such a determination cannot be made by scientists alone but, rather, should be open to debate among informed publics in a democratic society.[38]

On the face of things, this recommendation makes sense. In an era of artificial intelligence and biotechnology, society should have a say in what entails acceptable risk. But since the publication of *Risk Society*, the question has been what "having a say" should look like. As science and technology increasingly grow more and more complex and technical, what does greater public involvement in assessing the costs and benefits of science and technology entail?

One approach is to consider more carefully who should be involved in conversations about research risks and harms. For example, the Civic Laboratory for Environmental Action Research (CLEAR) is an anticolonial science lab based in Newfoundland and Labrador, Canada. From getting permissions to developing laboratory protocols to disposing of biological waste, the lab director and members reflect on every step of the process as a way of doing science that enacts ethical ways of being with human and nonhuman communities on the island of Newfoundland and beyond.[39] Because of these considerations, they create additional spaces beyond the standard institutional ethical review processes to allow communities to say no to certain research.

For example, CLEAR founder and director Max Liboiron and colleagues developed a system of community peer review after they first discovered plastic in the gastrointestinal tracts of cod. Although from a scientific perspective they were excited by the discovery, they knew that it could have profound economic, cultural, and social implications for Newfoundland. Earlier ecological research had led to a moratorium on cod fishing decades prior, causing many people to lose their livelihoods. The industry had still not recovered, and now a discovery of plastic could threaten the slowly made gains. Liboiron wrote of the story: "I remember staring at the little plastic fragment on my finger and thinking, 'How am I going to handle this?!' What are my obligations? How do I not cause harm? Then I thought, 'How would I know? I have to ask Newfoundlanders.' CLEAR's community peer review process was born in that moment."[40]

CLEAR's method of community peer review includes processes of consent, topic, and method development, not just whether it is acceptable to disseminate the results through publishing and other approaches. Liboiron and colleagues generously provide a step-by-step guide that can be adapted to other disciplines and contexts in their paper on the process, which would make excellent reading for any lab group.[41] This process should not be misunderstood only as a means to an end, however. There is a danger of seeking consent with the sole aim of gaining access or approval rather than with the spirit of seeing a nonacademic partner as having an essential, and equal, voice, and sometimes that voice will say no. During an interview, Liboiron equated this process to seeking consent for someone to go home with you from a bar.[42] Sometimes you might get turned down, and that's not just a refusal, it is an opportunity for the other person to affirm what they do want (perhaps to go home alone). As Liboiron wrote: "Refusal is affirmation and repair more than denial (though it's certainly that, too!). Refusal 'is not just a "no," but a redirection to ideas otherwise unacknowledged or unquestioned.' It can highlight and address the strained relationships between academics and communities, realign research values to local needs, benefits, and protocols, and, of course, bring attention to how the right to research is a colonial concept."[43]

As Liboiron and others point out, processes of getting permission are thus not just about avoiding harm but, rather, about being open to input to "do something else." In other words, sometimes choice in research means *not* doing something. Furthermore, processes of consent can offer the opportunity for additional input into one's research. When we tell someone about what we wish to do, we are not just letting them know of our intent, but providing an opportunity for thinking about whether it is ethical, doable, and potentially useful. Such processes can feel scary, and rightfully so, because true consent in research means the possibility of a full stop. But stops may be more like important pauses for reflection; they can be about "redirection" or the acknowledgment that it isn't the right place or the right time (or the right researcher) for a given research project. Seeing consent as a conversation and negotiation might mean the beginning of a fruitful research relationship down the road.

Another benefit to seeing processes of permissions and consent is as an opportunity to change broader institutional practices. For example, perhaps one of the most fraught types of consent comes from biomedical research on fetal tissue, which can require the consent of parents who have recently lost an unborn child due to miscarriage. The consent must be obtained within a very short window of time (approximately thirty-two hours) after miscarriage has occurred, which is a deeply vulnerable time for grieving parents. In addition, some of the uses of such fetal tissue may be disturbing to outsiders, as the process may involve transplantation into rodent hosts. However, these studies are seen by some as essential to advancing medical science and practice; fetal tissue is used to help people with spinal cord injury and degenerative eye disease, and some research involving fetal tissue can further the understanding of some of the environmentally induced diseases that contribute to miscarriages in the first place.[44] Thus, while donating fetal tissue provides no direct benefit to donors, the research that emerges from such donations can help save lives, including those of future unborn children.

Typically, such consent processes utilize blanket waivers, which give permission to researchers to use the tissue for anything they wish to do, without providing detailed information to donors. But this blanket waiver practice is increasingly being rethought for fetal tissue (and other biomedical) donations. The well-known story of the "immortal cells" taken without consent from Henrietta Lacks, an African American patient who died in 1951 of aggressive cervical cancer, has led to greater calls for increased knowledge and consent for the use of biological cells for research.[45] Taking on this call, a team of researchers with the Community Outreach and Translation Core of Northeastern University's Children's Environmental Health Center sought to create opportunities for scientists to reflect on how their consent practices would influence how parents donating fetal tissue would react to their research, a process they call reflexive research ethics. The discussions took place over two years, during which time the scientists debated both the benefits and challenges of providing parents with detailed informed consent, how much detail to provide, and potential unintended consequences or misunderstandings that could arise from such discussions. At the end of the process, the research team decided to switch from blanket consent to detailed consent as the standard policy for their research center.[46]

This example points to the importance of seeing consent as seen as part of a social process laden with relationships and consequences rather than just one that is about a yes-or-no ticking of a box. I often think of this example when colleagues of mine tell me of ways they attempt to circumvent research permissions protocols because they are time-consuming or bureaucratic. Indeed, they are these things and more, but they are also important spaces of opportunity, reflection, and communication. Perhaps all that will happen if you ask for consent is that you can tick that box, but as shown above, the practice of getting permission can also be a transformative, impactful activity.

Methods, Materials, and Protocols

Methods seem straightforward. We read papers in our field, we get advice from our advisors and colleagues, we check our budget, and voilà, we have our methods. For students learning about research, methods often seem dictated by—and limited to—best practices from our discipline. Certainly, that's an essential consideration. We cannot ascertain the amount of biomass in a wetland marsh by doing semistructured interviews with park users, just as we should not try to assess the cultural values of urban environments by using household census data. Any researcher worth their salt must have a solid handle of what types of methodological approaches are appropriate for a given research question, as well as their specific limitations and biases. But once we have determined a general methodology, there are further choices to be made. And these more nuanced decisions can lead us in interesting and novel directions.

One choice has to do with cost. Depending on one's field of study, costs for materials, equipment, and participants can vary from virtually nothing (archival research) into the tens of millions of dollars (pharmaceutical trials). It can often seem that our abilities to conduct research are determined by the amount of grant funding we are able to obtain. Get that big grant, we can do the research; fail to do so, we're out of luck. But there is another way to think about funding, and the lack thereof, that can potentially open up new ways of doing research—perhaps not for all areas of study, but for many. This approach, known largely as frugal science, refers to approaches to create and promote methods that are low cost and accessible for researchers and practitioners with fewer resources. For example, microfluidics—a type of portable and easy-to-use medical diagnostic device—have the potential to be an important tool in the early, accurate, and low-cost detection of infectious diseases across the Global South. But microfluidics requires the creation of a master mold, which is often made using procedures that are cost-prohibitive in countries with limited health care resources. An international team of researchers has begun to address this problem by developing methods using readily available and low-cost materials, such as nail polish, for the creation of master molds that bring initial expenses down from hundreds of thousands of dollars to about six thousand dollars.[47] Such approaches can help to democratize science and increase global equity not in just the products, but the process, of doing research.

Frugal science is central to the mission of Dr. Manu Prakash's bioengineering lab at Stanford University, where students from diverse disciplines and countries innovate to create research tools that can be produced with minimal resources for a multitude of purposes. Prakash's team's most well-known invention is a "foldscope," a fifty-cent microscope made mostly of paper that can be folded like origami.[48] Since the development and piloting of the innovation in 2014, more than 1.7 million foldscopes have been distributed across 135-plus countries.[49] The portable microscopes have been used for many types of research and applied uses, including to study fungal infections in agricultural crops, to quantify bacteria in wastewater, and in the detection of cancer cells.[50]

Another way to think about choice in methods is in terms of who is involved. Involving nonacademic partners in developing methods can lead to improvements, as additional perspectives can provide different insights, things that may not have occurred to us otherwise. One great example comes from a study conducted in 1996 of sidewalk concentrations of diesel exhaust particles in different parts of Harlem in New York City. Particulate pollution contributes to the incidence of asthma, which is a major health problem in the region, affecting more than one in four children (among the highest rates in the United States).[51] The study was conducted as a collaboration between researchers at the Columbia University Center for Children's Environmental Health and staff from West Harlem Environmental ACTion, Inc. (WE ACT), an environmental justice advocacy organization that serves an area of New York City known for its African American and Latino populations and rich cultural history.

Because WE ACT was deeply invested in the findings of the study, the organization's staff was engaged throughout the process of the research, including in the logistics of the methods. For example, initially, the scientists planned to place air monitors solely on school rooftops, a decision questioned by WE ACT community researchers involved in the project. They argued that the monitors should be placed in locations where children breathe, such as outside the windows of school buildings. This small adjustment to the methodology improved the accuracy of the research, as exposure to particulate pollution within a neighborhood can vary greatly depending on exact proximity to the source. Subsequent studies have demonstrated that living on lower floors of apartment buildings (and thus closer to street-level pollution) can directly influence indoor concentrations of street-level pollutants.[52] One of the researchers on the study later commented: "Sometimes as scientists we make assumptions and don't rethink assumptions to see how they fit in a natural situation. I think community (members), because they are looking at it from a fresh perspective, will question the assumptions in a way that actually improves the science. It may tailor things to the situation in a way we would not have thought of."[53]

WE ACT staff used the findings of the study to successfully advocate for major changes by New York City's public transit authority with regard to its bus fleet. For example, five of seven public bus depots (where buses are maintained while not in use) are in northern Manhattan. The study found that such depots were a major source of particulate matter pollution and advocated that the dirty diesel fuel be replaced with cleaner alternatives. Over time, WE ACT has been instrumental in the conversion of NYC's entire bus fleet from diesel, to compressed natural gas, and currently to electric.[54] Thanks to WE ACT's community-based researchers and activists, all New Yorkers now breathe cleaner air, demonstrating the power of place-based research to have impacts that extend far beyond the specific location of the research.

Another example of the importance of lay expertise in the design of research is in how HIV and AIDS patients and activists spurred rapid advances in medical treatment of the disease during the late 1980s and 1990s. Activists argued that most clinical trials, which were mostly made up of middle-class White men, were not representative of the different types of people affected by the AIDS epidemic. They advocated for clinical trials to be more inclusive of all social groups, including women, people of color, and those suffering from other illnesses.[55] Furthermore, they challenged conventional "best practices" used by the medical research community for setting up clinical trials, which excluded people who had already tried other treatments or who were taking unsanctioned medications. These actions changed the narrative in many conversations about research ethics that "doing no harm" also means giving people the right to make decisions about what harm means to them. As sociologist Steven Epstein wrote, "Most debates about the ethics of clinical trials in the United States in the last quarter century have focused on issues of informed consent and the right of the human subject to be protected from undue risk. AIDS activism has shifted the discourse to emphasize the right of the human subject to assume the risks inherent in testing therapies of unknown benefit and, indeed, to become a full fledged partner in the experimental process."[56] In this sense, the engagement of the patient-activists in the research process shifted not only how clinical trials were done in the case of AIDS, but had broader implications for research ethics and biomedical practices more generally.

No More Parachutes

There are other ways to think about choice when setting up a research project, including additional questions of ethics, materials, and personnel. Just knowing that there are countless choices not just in what we do, but how we do it, can help make science more accessible, equitable, and less harmful. What many of the examples in this chapter suggest is that enabling others to weigh in on some of the decisions made in research can be a powerful choice in itself. As science communicator Faith Kearns pointed out, for many groups that have been underrepresented in science, "access to information is not as big a barrier as access to deliberation."[57] This point is essential because participation in deliberation can be transformative. For example, although I first went to Nicaragua as a parachute scientist, I did not stay one for long, largely because the community where I did my research did not allow me to be one. I was fortunate to connect with people who guided me in more equitable ways of doing research (as I didn't get much of this training in academic institutions in the United States or England). People took on the labor of educating me on what I should have known before I arrived. They conveyed the importance of hiring locally, being respectful of cultural norms, getting consent at multiple levels, and returning the results of research in a way that was useful to community members. I was a parachute scientist who thus learned some different and better ways of doing research, mostly to the credit of those from outside academia.

That said, training students to do equitable and ethical research should be the responsibility of the academic community, not community members in over-researched parts of the Global South. Thankfully, things are beginning to change. For example, due to recognition of the central role played by academic journals in determining what research is published, editorial boards of some prominent journals have begun taking steps to promote fairness and equity in global research. *The Lancet*, for example, rejects "papers with data from Africa that fail to acknowledge African collaborators."[58] And some natural history museums have instituted requirements for researchers to deposit specimens at local institutions; similarly, some countries, such as Panama, have also developed protocols that foreign researchers are required to follow.[59] Many scientific societies also have a code of ethics to encourage collaboration and benefit sharing in research; however, they often lack means of sanctioning researchers who do not abide by them.[60] Thus, while the academic community develops appropriate ethical guidelines and the means of implementing them, the choices individual researchers make while designing their research projects will still largely determine the impacts of research in a given place.[61]

Finally, sometimes choice in science is about *not* doing something. For example, in 2011, climate scientist Peter Kalmus decided to stop flying because of the negative impacts of airplane travel for global greenhouse emissions. Kalmus was a postdoctoral student at the time, and the decision initially had some negative impacts for his scientific career as it limited his ability to travel to important science conferences. But he also found that it made him more available for local and regional opportunities and honed his focus in terms of the type of scientist, and person, he wanted to be. He now credits his decision to stop flying as one of the most important of his life: "I stopped taking food, water, air, fuel, electricity, clothing, community, and biodiversity for granted.... Now, I feel more connected to the world around me, and I see that fossil fuels actually stood in the way of realizing those connections.... By changing ourselves in more than merely incremental ways, I believe we contribute to opening social and political space for large-scale change."[62]

As Liboiron expressed to me, "Sometimes you don't have to do something else. Sometimes not doing it, saying 'Maybe, just not,' is the most powerful thing that's in your repertoire, in your jurisdiction, and the most important impact that you have available right now."[63] There is a saying that anyone who has had their heart broken, lost their job, or been rejected from a seemingly great opportunity has heard, "When one door closes, another opens." By bringing others into the process of determining risks and harms and by being more conscious about the choices we make while engaging in research, we can have a more democratic process of choosing which doors should be closed, even if just temporarily, and which should be opened.

CHAPTER 6

The Power of Participation: Data Collection and Analysis

I CAN REMEMBER THE EXACT MOMENT I DISCOVERED my love for science. The year was 2006, and I was in a small farming community called El Arenal, Nicaragua, doing research for my master's thesis. My topic was broadly about understanding local perceptions on environmental issues in the community, which meant using social science methods such as interviews, observation, and workshops. I learned a lot from these qualitative approaches, but I was also interested in describing the ecological impacts of land use in the community more quantitatively.

One morning, my research assistant, Donal Pérez Gutiérrez, and I grabbed our notebooks and a measuring tape and walked along the main road of the community. Although we walked along the road almost every day, that morning Pérez and I had a specific purpose. We were there to conduct a very small scientific study, one that would allow us to answer a simple question about whether one area of the community was more biodiverse in tree species as compared to another. For this exercise, we had laid out a hundred-meter transect along the road, and at intervals of twenty meters we measured and identified the trees within a specific distance from the road. Later, we would plug the data into an Excel spreadsheet and use a mathematical formula that had been created for this purpose to answer our question.

In many ways, I was an unlikely person to be measuring trees and calculating indices. Science never came easily to me. As a teenager, I was bored by being forced to memorize concepts that seemed increasingly abstract the more I advanced in my education. From parts of cells invisible to the naked eye to the order of elements on the periodic table to abstract equations with predetermined answers, science seemed to me rote, predictable, and unrelated to things that interested me. And perhaps in part because of my lack of interest, I wasn't particularly good at it. Biology was manageable, chemistry was a struggle, and physics was a disaster. I distinctly remember chasing my high school physics teacher through the parking lot as he rushed to his car, so desperate was I to hear if I had passed the state exam (I had not).

So, when I went to college, majoring in a science field was not even within the realm of possibility. After fulfilling the sole basic STEM course required for my liberal arts degree, I thought science was out of my life forever.

But there I was, less than a decade later, doing my own small scientific study with Pérez, my friend and research assistant. Pérez was from just a few miles away, where he grew up on a farm in a family with ten brothers and sisters. He was passionate about trees and plants, and at the time, was in the process of completing a degree in agricultural and forestry systems at the Universidad Nacional Agraria in the capital of Managua while at the same time helping his father with the family farm. Assisting me was a kind of third job, yet one that brought in useful income in a place where cash was often hard to come by.

Earlier that week, Pérez and I had done the same measurements in another part of the community, one where there were no homes but where community members frequently went to collect firewood. Because that area of the community was also accessible to loggers, we hypothesized that the tree diversity would be greater along the road, where people had a strong interest in having shade and fruit trees. Indeed, as it later turned out, after I had crunched the numbers, the data supported our hypothesis. But getting an expected result was not what was interesting about the exercise. Rather, it was the conversations.

By that point in my stay in El Arenal, I had had countless conversations about my research. I had spent the previous months sitting in people's living rooms and explaining my thesis, asking questions about environmental and other local issues, and talking about difficult concepts such as sustainability and climate change. Hundreds of conversations later, I thought I had figured out the formula for a good chat about local environmental issues. But the conversations that Pérez and I had with the neighbors along the road that day were different than any I'd had before.

As the trees were mostly on private property, at almost every house along the road we had to stop and ask for permission to measure a tree and explain what we were doing. Because the method was standardized, we couldn't only stop at the homes where we were friendly with the owners; we had to interact with whoever happened to live every twenty meters. Although some people were indifferent to our request, shrugging their shoulders to indicate that yes, it was fine, or, in a couple of cases, saying no, most had questions for us. What were we doing? Why? What did we expect to learn? How was this information useful? Did it tell us something about their land? Was this exercise related to the thing they had heard about called climate change? More often than not, they accompanied us to the tree to watch us measure it and told us about its history before we moved on. Furthermore, we began to pick up additional helpers along the way. Whereas at the start it was just Pérez and me, by the end of the transect we had two other young people join us, both of whom expressed interest in potentially studying environmental issues at the university. Having just spent the previous few months trying to manufacture such discussions in people's living rooms, I was amazed. What was the power of this approach in generating interest, follow-up questions, and even future environmentalists? To my great surprise, I realized it was science.

Science as a Verb, Not as a Noun

One of the biggest misconceptions about science is that it is a collection of facts rather than a process of inquiry. In other words, people tend to think of science as something that one can come to "know" rather than something that one can "do," largely because that's what is often taught in school. Less engaging, conventional ("sage on the stage") science education uses the "fill your brain of facts" approach familiar to many of us. Such knowledge is easier to test but has the unfortunate consequence of teaching students that science is about learning already established answers rather than seeking out new knowledge for oneself. Knowledge not directly related to addressing questions of significant interest or not used to solve a problem is soon forgotten. In an article on rethinking science education, Bruce Alberts, biochemist and past president of the National Academy of Sciences, relayed the following anecdote:

> A scientist parent notices that her elementary school child has thus far not been exposed to any science in school. As a volunteer teacher, she begins a science lesson by giving the children samples of three different types of soil. Each child is told to use a magnifying glass to examine the soils and write down what they observe in each sample. She waits patiently, but the children are unwilling to write anything. Her probing reveals that after three years of schooling, the students are afraid to express their views because they don't know "the right answer."[1]

This sense that science education is about teaching students "the right answer" persists at all levels and in most countries around the world. It was the main reason I always believed I didn't like science and I wasn't good at it. A quick look at the statistics shows that I was not alone in thinking that science wasn't for me. Research has found that although many young people are interested in science, such interest does not often translate into aspirations to pursue science-related careers.[2] Further research has found that even students who decide to major in science in college often switch majors in their first or second year after struggling with lecture-based required courses that are often designed to "cull" students who might struggle with the material.[3] I've had multiple advisees tell me that they dropped biology or chemistry as a major when their professors announced on the first day of class that a certain percentage of students were likely to fail. Such negative experiences with science in formal educational settings have been found to have particularly adverse consequences for students from communities that have been underrepresented in science, particularly ethnic or racial minorities, low-income students, women, and students from rural backgrounds.[4]

One way to counteract the widespread lack of connection that many people have to science and scientists is to increase a person's "science capital," the combined science-related experiences and knowledge a person may have, including connections with others who are working in science fields.[5] Research has found that young people with higher levels of science capital have been found to be more likely to maintain science-related aspirations than those with lower levels of such capital.[6] Broadly speaking, increasing science capital is about creating more science-based experiences for people, such as opportunities to participate in out-of-school science activities, meet people in science-related careers, and have conversations about science.[7] Although more traditional science literacy approaches (such as mandated science classes) can also contribute to one's science capital, focusing on creating positive science-related experiences can foster greater connection to science.[8] Most of us who teach science know that students learn better when they are doing rather than just copying down information. We can give them a lecture on the ecosystems of secondary forests, and they may learn enough to pass a quiz on ecological succession, but if we take them on a field trip to a local forest, they get a broader and deeper sense of the key concepts and issues. They might hear the birdsong of the wood thrush and learn about the types of species present and thus understand the value of such places in a way that would not be possible through the pages of a textbook. They might come across the remnants of a foundation of a barn and thus learn the agricultural history of the land and begin to question the premise that humans are separate from nature.

That's the whole basis of experiential education, which is an increasingly recognized high-impact pedagogical practice. Experiential education is based on the premise that we learn best through direct experience and engagement, which often involves leaving the classroom, engaging in new experiences, and then reflecting on those experiences. Although theories of how experiential learning works have been around for decades, not until recently have rigorous assessments (such as randomized control trials) been conducted to assess student outcomes.[9] Early research has been promising, and experiential learning has been shown not only to increase interest in a given topic, but also performance.[10] Such approaches can be particularly powerful when they are done in ways that are relevant to a person's life experiences. For example, I AM STEM is an organization based in Atlanta, Georgia, that uses experiential learning, such as science experiments and field trips, to engage students from communities that have been underrepresented in the STEM workforce in science.[11] Founded by science educator and researcher Natalie S. King, the program was developed to be a "counterspace" in which women of color could challenge misrepresentations about their abilities in STEM and "fully engage in experiential learning and lessons that directly connected science with their interests, lived experiences, and local concerns."[12]

Connected to the idea of experiential learning is that of embodied cognition. Whereas experiential learning largely developed from pedagogical theory and more qualitative educational studies, embodied cognition stems from psychological theories about how information is processed in the brain. Scholars of embodied cognition suggest that information is grounded in both perception and action and that the learning process is deeply associated with the physical experience of the learner.[13] Because learning and the processing of information are taxing tasks, our brains have developed ways to reduce the effort involved, and one of these strategies is by thinking with our bodies.[14] In other words, we store information in our senses and in cues that we get from our physical environment. Physically enacting something can also lead to increased recall, which is another reason embodied learning is so powerful. Multiple studies have demonstrated the value of handwriting notes over typing them out, something that one of my colleagues likes to point out to undergraduate students at the beginning of every semester. Because the act of moving the pen requires more precision than is required for the pressing of a keystroke on a computer, taking notes or copying passages by hand consistently leads not only to greater recall of information, but, impressively, is additionally correlated with improved learning.[15] Embodied learning can be especially powerful for children and for people with learning differences.[16]

What these educational and psychological theories all suggest is that there is great power in providing opportunities for people to physically "do" science. Seeing science as a verb rather than a noun emphasizes the dynamic and active nature of the scientific process instead of as a static collection of facts or knowledge. And the stage that offers the most potential for this "doing" is usually data collection. In fact, recent research has found that for increases in scientific literacy and understanding, engagement in the data collection stage of research is more important than any other stage of the scientific process.[17]

Participation in Data Collection

That day of measuring trees with Pérez was the beginning of a new way of thinking for me, a new interest in something I thought was out of my life forever. On that walk I got a taste of something that, despite more than twelve years of mandated science education, I had never before encountered: science wasn't just about memorizing concepts or performing experiments to which my teacher already knew the answer; rather, first and foremost, science was about being curious and asking questions. Perhaps even more surprisingly, I realized science was something that one could do with other people. It was social.

Immediately, I was hooked. In a stroke of good fortune, after completing my master's program and searching for jobs, I came across an incredible opportunity to assist on an ecological research project. The project was based in the New York metropolitan area, not the place most people associate with ecological research, but over the next few months there, I learned more about science than I had in all my formal years of education combined. The project was unique in that we were studying the abundance and diversity of multiple taxa (plants, mammals, birds, and amphibians) in sites located in the heart of New York City as well as in protected forests in the rural parts of the state. The goal of the research was to better understand the impacts of urbanization and land use change on different species and whether specific habitat types were particularly important for supporting biodiversity.[18] During the project, I learned to identify more than thirty bird calls, set up small mammal and camera traps, and spot amphibian egg sacks in vernal pools.

At the project's end, I was still curious to learn more about the local ecology of the region, so I teamed up with wildlife biologists Mark Weckel and Chris Nagy to conduct New York City's first study of urban coyotes, which eventually blossomed into the Gotham Coyote Project.[19] The arrival of coyotes to New York City is a very recent phenomenon with important ecological and resource management implications for scientists and park officials, and our project was thus designed to determine the presence or absence of coyotes in parks in the region.

Both projects were funded by the Earthwatch Institute, an organization that provides scientific research experiences for paying volunteers who come along on the "expedition" to help with data collection.[20] The work wasn't exactly glamorous; to the contrary, a big part of the coyote research involved putting out smelly bait disks and looking for scat in overgrown urban parks, where the biggest dangers were ticks and poison ivy. The project did, however, offer the opportunity for the volunteers to learn about coyote biology and the natural history of the region, as well as other local opportunities to get involved in protecting wildlife.

The volunteers, who were mostly retirees, were enthusiastic. Some of them stayed in touch long after the project ended, and one couple continued to collect and store coyote scat in their freezer. I began to wonder what specifically they were getting out of the experience. Was helping with data collection leading the participants to care more about urban wildlife? What about conservation attitudes? Did it cause them to engage in more environmental advocacy? What about science more generally?

To answer some of these questions, I created a short survey to ask the participants about their experiences. The results suggested that although most of the volunteers had come in with prior interest in environmental issues, their experiences in the project had led them to think more positively about wildlife and to engage in additional activities to support their local environment.[21] For example, one participant (a teacher who had gotten a scholarship to attend the expedition) said: "Participating in the project made me feel energized and excited about (local) conservation. I learned a lot about coyotes and native plant species, which has influenced my curriculum and decisions I make in teaching. We have a planting program at school and since participating in the coyote project I have advocated for planting indigenous species of plants to support wildlife in the area.... The three-day project made a big impact on me."[22]

These results were self-reported gains, so I couldn't say for certain that the perceived impact of the experience wasn't just a way for the volunteers (with whom I had spent three days) to give me positive feedback about an experience they had enjoyed. But I found it interesting that people would say that engaging in the research project had led them to both think and act differently, and I wondered why that was.

The Relationship between Thinking and Action

Unbeknown to me at the time, there was a massive amount of scholarship on the questions in which I was interested, particularly coming out of the fields of psychology and education. For decades, scholars in these disciplines have been seeking to understand why some people act in "proenvironmental" or "prosocial" ways. Why do some people choose, for example, to become vegetarians or donate to charities or vote for ballot measures that would raise their own taxes but support social programs? On one hand, it was clear that people's attitudes informed their behaviors. In other words, if someone felt strongly about a particular issue, they were more willing to act to support or defend it. But on the other hand, people often act in ways contrary to their beliefs. For example, a person could be passionately outspoken about climate change yet also live in a massive home in the suburbs and drive a luxury gas-guzzling SUV.

Theoretical models of the relationship between attitudes and behaviors thus began incorporating additional social, economic, and psychological factors to explain how they can act as either barriers or incentives to behavioral change.[23] Interestingly, some approaches suggested that the relationship between attitudes and behaviors was bidirectional; in other words, it wasn't just attitudes that could influence behaviors, but that behaviors could lead to changed attitudes.[24] A theory I found particularly compelling was self-perception theory, first proposed by social psychologist Daryl Bem in the late 1960s, which suggests that we come to "know" how we think and feel about something in large part by observing our own behaviors in relation to that thing.[25]

A powerful example of this theory comes from a fascinating study that sought to encourage condom use among US college students during the AIDS epidemic in the early 1990s.[26] Researchers were finding that neither fact- nor fear-based messaging about AIDS was very effective for increasing condom use. Although most young people believed that AIDS was a serious problem, they did not necessarily see it as being *their* problem. In this study, college-aged research participants were told that they would be helping with the development of an AIDS prevention program for high school students and were asked to record televised messages explaining why condoms were important. The researchers found that intent to use condoms greatly increased among those who "preached" as compared to those who were just asked to reflect on past behavior. In other words, engaging in a behavior that supported condom use led participants to realize that they in fact did believe that condom use was also important for them.

Self-perception theory helps explain why engaging in an action—not just thinking about it—can be so powerful. We often don't really know how we feel about something until we try it. When we observe our own behaviors related to a given topic or issue, we develop a better understanding not only of our feelings about that issue, but more powerfully, of ourselves. That's why doing the tree transect with my friend was such an important turning point for me in how I perceived science and my ability to do and like doing science. Self-perception theory also helps explain why the volunteers on the coyote research said that the project had caused them to think and act differently about other local conservation issues.[27] This finding seemed incredibly powerful to me, as it suggested that instead of trying to get people to *think* differently about science or science-related issues, such as climate change, it could be more effective to create opportunities for people to participate in doing science.[28] So much science communication and education are focused on shifting peoples' opinions or attitudes. But what if we focused on providing positive experiences for people to engage with science instead?

The Many Forms of Participatory Science

Creating opportunities for laypeople to participate in scientific research has a long history and is often referred as citizen science.[29] In a nutshell, citizen science is an umbrella term that refers to the participation of nonscientists in the process of scientific research, most frequently during the data collection stage. Citizen science has exploded in popularity, with thousands of projects engaging millions of volunteers, including opportunities to contribute to medical research, engage with astrophysical data to observe exoplanets, and study the effects of climate change on biodiversity and local environments. Such projects build on the educational and psychological theories mentioned above, providing needed opportunities for people of all ages to learn that science is something that one does, not just something that one knows. Furthermore, participation in the process of doing science can lead people to question or reinforce their values associated with the topic being studied.[30]

Citizen science projects often have a broad range of aims; some are more educational, while others seek primarily to gather data that would otherwise be cost or time prohibitive. For example, Zooniverse is an online platform where researchers can upload data in need of classification, and citizen science projects on the Zooniverse platform include categorizing superluminous supernovae, counting deep ocean fish, identifying histories of incarcerated individuals, and looking for patterns in lymph nodes of breast cancer patients.[31] Such projects have thousands of volunteers, and some have led to important scientific discoveries. For example, in 2017, citizen scientists helped discover a rare brown dwarf (a substellar object that is larger than a planet but smaller than a star) while participating in the Zooniverse Backyard Worlds project.[32] Many projects also enable discussion and collaboration among enthusiasts via online discussion boards, where volunteers can post questions and discuss unexpected findings with research teams.

Other types of citizen science projects aim to enhance scientific and public understanding on issues that have been understudied. For example, the winner of the 2023 European Union Prize for Citizen Science was a project created by researchers at the University of Antwerp in Belgium that seeks to develop a better understanding of the female microbiome. Isala, as the project is called, was named after Isala Van Diest, the first female doctor in Belgium and a prominent activist for women's health in the nineteenth century. Volunteers in the project self-collect vaginal swabs in the privacy of their own bathrooms and then send the samples to the researchers for testing. Within one year of submitting their samples, each participant receives their personal vaginal microbiome profile. The project not only aims to improve scientific knowledge of the female microbiome, but also has important societal and health goals, such as breaking taboos around vaginal health, and thus far has engaged thousands of women.[33]

Other projects seek to generate data for the explicit purpose of influencing or informing policy. For example, the South Atlantic Fishery Management Council (SAFMC), located along the Eastern Seaboard of the United States, developed a citizen science program to fill data gaps in fisheries science and management. Julia Byrd, SAFMC's citizen science program manager, said that one motivation for the project was to build trust between researchers, fishers, and others involved in fisheries management. As Byrd told me, "We realized that it was really important for fishermen to be involved in making the decisions about their fisheries. Things that they were seeing on the water were things that may not necessarily show up in the data yet. And there was just a lot of knowledge there and also a fair amount of distrust between the people who are managing the fisheries and the people who are actively fishing on the water."[34]

Although many citizen science programs may hope their data will be used to inform policy but don't actively plan for it, SAFMC's citizen science program was designed with built-in data infrastructure to ensure that research would be both relevant and usable.[35] All citizen science projects developed under the framework are required to have the potential to contribute to that information base. One project that met this requirement was called SAFMC Release (housed in an interactive app called SciFish), where fishers can report data on released fish such as grouper and red snapper, species for which limited data was available to assess stocks. Fishers were actively involved in the development of the app to ensure its practicality both aboard boats and back at the dock and have helped recruit others to participate in the program. The project has attracted interest from other government agencies, as many fishery councils face the challenge of how to address data gaps associated with released fish. However, Byrd pointed out that these successes were hard-won and emphasized the importance of building relationships and community in the process: "People think that setting up a citizen science project is easy, and you set it up and have volunteers, so you can do it on the cheap. But that isn't the case. It's not a no-cost endeavor. The things that are taking most of our time are with volunteer engagement and recruitment and retention over time. I think people who don't work in this space think that it's like, 'Build it and they'll come. They'll use their app. Just tell 'em it's there.' That's not how it works."

What Byrd's comment points to is that running a successful citizen science program—particularly one that is designed to inform management—requires investment over the long term. Support is needed not only to build the infrastructure needed to collect and store data so that it can later be used, but also in terms of building relationships and a sense of community among participants. People must believe that the project is worth their time and effort, which can be challenging in policy landscapes where there is often a significant time lag between data collection and its application to management. In research that I have conducted on citizen science outcomes, I have found that most volunteers join to connect with others, contribute to science, and potentially "make a difference" in terms of affecting policy or practice.[36] But different types of participation can lead to different types of outcomes, so it's essential that researchers who seek to incorporate public engagement in their projects understand the motivations through which people come to participate, as well as the different levels at which they engage.[37] Regardless of the level of participation, one finding across all projects is that interaction between researchers and participants is a crucial element of what makes a project successful.[38]

Another stage in which people can be involved in "doing" science is in the interpretation of data, such as using qualitative analysis techniques. Sociologist Anne Byrne and other researchers developed a participatory research project with Irish teenagers whose negative experiences with formal schooling had led to them leaving prior to graduation.[39] The project sought to provide the young people with skills to examine their experiences on their own terms, which included training in the analysis of interview and focus group transcripts. Initially, the researchers were concerned that the teenagers would struggle with interpretation, which required generalizing beyond the specifics shared in the personal accounts. But to the contrary, they were not only able to do so, but noted important aspects of the research that the academics had left out. While the trained researchers focused their interpretation on how the teenagers' disadvantaged stories led them to leave school, the teenagers pointed out that the academics' privileged stories were essential for understanding their comparative success with schooling. They wrote: "Our analyses were duly amended to include the missing narratives. In erasing stories of educational advantage, what psychological processes were at work? Were we perhaps afraid of what differences (class, gender, age) among us might reveal?"[40]

As the quote above suggests, engaging stakeholders in this type of participatory process reveals key points of conflict and disagreement to how data is being interpreted. That can be both enlightening and anxiety-inducing for researchers, who must find ways to negotiate between their "outsider" perspective and that of participants, for whom the research may feel like an intimate exploration into their lives.

Hundreds more projects could be mentioned here—a quick search on the citizen science website Scistarter gives a sense of both the scope and depth of opportunities available to engage in the data collection and analysis stages of research. But as Byrd pointed out, the wide range of existing programs does not mean this work is easy. Developing a citizen science program that results in robust data, opportunities for education, or policy outcomes can take years, particularly if the creator seeks to achieve multiple outcomes. It is essential to be realistic about time frames and the extent to which resulting data can be used (whether for scientific or applied purposes). Providing volunteers pathways to transition from being data collectors to doing something more with the data is not straightforward. Thus citizen science projects need to be an integral part of overall project design and purpose, not just an add-on to attract funders.

Furthermore, an additional consideration is *who* is being engaged by citizen science projects. Studies of demographics of citizen scientists across Europe and North America have consistently found that they are overwhelmingly White, with a college degree, and with above-average income.[41] Changing this landscape requires a willingness on the part of researchers to listen to the concerns and interests of communities that have been historically underrepresented in science and create projects to reflect those concerns. Environmental justice organizations have been doing this work for decades.[42] For example, the West Oakland Environmental Indicators Project is a nonprofit organization based in a predominately low-income Black and Latino community in California exposed to environmental harms from local industrial sites and freeways.[43] The organization conducts community-based research on issues of local concern; residents participate in the development of research questions, data collection, analysis, and the use of data to advocate for legislative and regulatory policy in Oakland and beyond. The organization emphasizes the importance of long-term, sustained engagement for policy outcomes that reduce inequities. As stated on its website: "Community power in the 21st century must be backed up by community knowledge. That knowledge depends on strong research partnerships involving those impacted by the data."[44]

Making Time and Space for Unexpected Conversations

Not all research projects have the budget or the capacity to include laypeople in the data collection or analysis stages of the research. Data collection and analysis are time-consuming, iterative, and often require certain types of expertise, such as knowledge of statistics or the ability to "code"—whether qualitative data or in stats programs such as Rstudio. Data can also be rife with confidentiality issues, especially with regard to medical or social science information. But that doesn't mean there aren't opportunities for impact during these stages. All research, whether conducted in the field or the lab, is happening on somebody's land, and most is conducted near places inhabited by humans. Thus there is potential for interactions with the public, whether scientists seek them or not.

My friend and colleague Monica Palta is an expert at making the most of these types of encounters. She is a wetland ecologist whose work often requires her to walk through polluted waterfront areas, taking water samples and measuring the height of marsh reeds. Because her field sites are typically in urban or suburban areas, she regularly gets questions from people—usually about whether the waterbody she's studying is polluted or not. On occasion, I've accompanied her on this work and have marveled at her patience in these situations. Palta never treats anyone from the public like an unwelcome interruption. She pauses what she's doing to give the person her focused attention, allowing them to ask any questions they like and using the opportunity to ask questions of her own. Sometimes these conversations will last for fifteen, twenty, or even fifty minutes, but they only end when the other person is ready for it to end.

Palta told me that she largely learned this approach from her father, Jiwan Palta, emeritus professor in the Department of Plant and Agroecosystem Sciences at the University of Wisconsin–Madison. The senior Palta grew up in a rural part of northern India where farming was an important activity. He trained as an agricultural scientist in India and then came to the United States to pursue a master's in soil science and then a PhD in plant cell biology. When he was first hired as an assistant professor to conduct basic research in cell biology at Wisconsin, he knew he wanted to do applied research geared to solving problems in production agriculture, so he sought out connections with local farmers. Making connections was challenging at first; the predominantly White cranberry and potato farmers were not used to meeting people from other cultures. Furthermore, although Palta had conducted basic research on potato plants, he knew little to nothing about cranberry crops. But he was interested in the farmers and their ideas and built his research program in part to address the issues that most mattered to these farmers, similar to the approach taken by John Tooker in chapter 4. As Palta explained to me:

I had not seen a cranberry before in my life before I came to Wisconsin. I didn't know what kind of plant it was, but I learned so much from the farmers. Because there are not many scientists working on cranberry, cranberry farmers have to depend a lot on themselves. They have been very innovative and devise techniques, how to apply fertilizer, how to control weeds, how to manage the crop. So I learned a lot from them. Then I asked, okay, so what are your issues? That's when I started to connect my basic science and background to address the issues important to the industry through the knowledge I had learned from the farmers.[45]

The combination of crop science and farmer knowledge proved to be hugely impactful for cranberry and potato farmers. For example, because the threat of frost damage to the fruit in late fall is a major concern to the growers, one issue farmers were interested in was how to ripen the cranberry crop earlier in the season. Color and flavor development, which are very important to juice and fresh markets, occurs as fruit ripens naturally in response to cold temperatures. Ripening methods that worked on other crops, such as spraying the crop with a liquid form of manufactured ethylene (a fruit ripening agent), had not been found to be effective for cranberry ripening. Palta soon discovered that the cuticle of cranberry fruit is so thick and waxy that conventional water-based applications of the chemical were ineffective. With his background as a cell biologist, he made an important discovery: the cuticle would take in the ethylene if the formulation was mixed with lipids (oils) rather than water. Further experiments with farmers in their fields led to an even more surprising finding; the "control" plots that were sprayed with only the lipids (without the ethylene) were equally good at ripening the fruits! In other words, the chemical wasn't necessary.

This latter finding was transformative not just for cranberry science, but for other fruits as well, as the fruits ripened with lipids alone were found to have better shelf life since they did not soften in storage. Many fruits such as tomato and banana turn soft when they ripen and thus have poor shelf life. This latter finding was initially hugely controversial, as it challenged decades of what scientists thought they understood about how fruits ripen.[46]

Palta's research demonstrates the importance of farmer knowledge for informing not just applied but also basic science, and his daughter has followed in his footsteps. For example, the younger Palta told me that she views conversations that arise in the field as opportunities not just to share what she knows with the public, but also for her to gain important local knowledge of her field site, knowledge that is often not written up anywhere. For example, her extensive time speaking with people in the field has led to a better understanding of less conventional uses of urban waterways by members of marginalized communities, pointing to the broader value or potential harms associated with such spaces.[47] Although care must be taken to think through ethical issues of consent, such encounters in the field have the potential to be important opportunities for sharing ideas.

It takes a bit of reframing in our own minds to see interactions during the data collection stage as spaces of encounter and connection rather than interruptions. Part of the challenge is that they aren't the types of "public outreach approaches" easily described in grant writing or reporting. They are unplanned, unpredictable, and hard to quantify. Unlike a formal public event, such as a science café, we will not be able to control the conversation by, for example, having a designated time for presenting and then a limited follow-up time for audience questions, but such encounters can be valuable spaces for both engagement and impact.

Coda

It has been almost fifteen years since Donal Pérez and I measured trees along the road in El Arenal. We recently chatted on WhatsApp, and I asked him if he remembered that day and if it had been as powerful for him as it had been for me. He said, in his quiet way, "I would never forget it. The experience sparked in me an interest in continuing to work with trees."[48] "Of course," I said, remembering that toward the end of my stay in Nicaragua, I helped connect him with a job opportunity with an organization I was volunteering with at the time, called Trees for the Future. Pérez spent the next seven years planting thousands of trees in and around his community, particularly on farms like his family's. I later asked him if he would mind writing up a short reflection of what he remembered from our time working together so that I could include a quote from it in this book. He kindly did so, writing out a description of what we did from his perspective. One thing he wrote resonated with my own experience: "What I liked most about the work were the conversations that Anne and I had, as well as listening to what people in the community had to say."[49] Pérez has continued to listen and have conversations. Currently, he lives in El Arenal with his wife, son, and daughter and is a community leader for GRUDESA (which stands for Grupo de Solidaridad-Arenal, the Arenal Solidarity Group), a group of friends and neighbors that organizes and carries out educational, health, and environmental projects in the community.

CHAPTER 7

Rethinking the "Peer" in Peer Review

IN THE LATE 1990S, A SCANDAL EMERGED in the United Kingdom when one hospital, Bristol Royal Infirmary, was found to have low rates of survival for children undergoing heart surgery for congenital heart disease. According to one estimate, during a five-year period, more than thirty babies died in the unit who might have survived if treated elsewhere.[1] In the inquiry that followed, the National Congenital Heart Disease Audit was set up to collect data on postsurgery survival rates and make such data available to the public online.[2] Parents and clinicians of patients undergoing heart surgery were encouraged to regard the data as a useful source of information for making decisions on treatment options.[3]

However, the data that was collected and shared was both incomplete and difficult to interpret, thus limiting its use as a tool for accurate assessment and decision-making. For example, some units were well known for taking in the most difficult cases and thus had higher mortality rates than those that only dealt with less complicated cases. For a parent or journalist looking at the hospitals ranked by survival rate, this crucial detail was not apparent, thus leading to conclusions that the best hospitals were those with the highest survival rates. But, in fact, some of the top surgical units in the country were those with the lowest survival rates because they were taking on the most complex cases.

To address this gap in knowledge, researchers at the University College London developed a risk model in 2011 that would allow hospitals to adjust for the different case mix and be able to better evaluate how different units were doing. The resulting model, called Partial Risk Adjustment in Surgery (PRAiS), was designed alongside software that hospitals could use to keep track of surgery-related data and use the results for improved care within their units.[4] One of the challenges were gaps in reported data (such as the weights of patients), and thus a key aim of the accompanying software was to flag when crucial data were missing to ensure the accuracy of the model. For example, in one case, missing data led to the misinterpretation of survival rates and the temporary closure of one important surgical unit.[5] This closure was extremely distressing for parents whose children were being treated in the unit, some of whom had to then travel farther to find similar care. Researchers studying the data argued that it was being misunderstood and thus misapplied and that there was no good motivation for the closing of essential units. But something needed to be done to better explain the confusing statistics to families, journalists, and health care providers. As described by PRAiS lead, Christina Pagel:

> Seeing how it was covered in the press made me realize that people just didn't understand what the National Health Service were doing with that data. They didn't really understand what the risk model was. They didn't understand about how the audit worked, they didn't understand the implications of what the mortality rates were and so on. And then I realized there wasn't really anything there for parents to help them understand what any of this stuff meant. Even though the results are public, they weren't understandable to the non-expert person. . . . I

think until then I hadn't really thought about the communication side of it. And I guess what I would've said, thank you, it's not my responsibility how it gets used. I'm a mathematician, I come with model. But then after that I thought, well, if it isn't my responsibility, then whose responsibility is it?[6]

So, in 2015, Pagel and her colleagues teamed up with the organization Sense about Science to create a website to communicate the survival outcomes of different hospitals in the United Kingdom for young patients with congenital heart problems. For the lead researchers, who were primarily mathematicians and statisticians, doing this type of science communication was new—they were trained to work with numbers and create models, not in how to engage with people. In fact, the initial idea for the website was something of a "tack on" to a larger research project to update and improve the existing model used to calculate risk data.

But Sense about Science had a solid track record of effective public engagement in science. The mission of the organization was to challenge the misrepresentation of evidence in public life and call for the honest and open sharing of research at multiple levels. The team had more than a decade of experience creating opportunities for the public to engage with evidence in multiple ways, from reducing the mystery of clinical trials and peer review to creating spaces in which various publics can ask informed questions about science. Tracey Brown, director of Sense about Science, explained to me how the project started: "A group of statisticians came to us and said, 'We're doing some stuff with this formula about how to work out whether the outcomes of the hospitals are within the predicted range. And to show it in such a way that it's helpful to the parents.' And we said, 'Well, let's get the parents in now.'"[7]

In collaboration with the Children's Heart Federation, the team immediately brought in parents of children who were undergoing heart surgery, as well as journalists and clinicians, to help with the website's design. Through a series of workshops, participants reviewed multiple iterations of the data visualizations that had been created by the researchers, asked questions, and gave feedback about what was easily understood and what was confusing. For example, researchers needed to present the data in a way that avoided direct comparisons between different hospitals, as additional context was needed to understand what led to higher survival rates in one versus another. The researchers realized that rather than providing the exact predicted survival rate for each hospital (which encouraged direct comparisons), it would be more useful to provide the actual survival rate for each hospital within its predicted range (which more accurately described survival rates in relation to the types of patients being treated). As Brown explained:

> The parents just took over. I mean, they redesigned everything and you ended up with them re-explaining back what an expected range was, what uncertainty was. This is all from them, and they'd designed the colors, they switched it all around, so you didn't talk about mortality rates, you talked about survival rates. So it became a lot easier to spend time on the site because it wasn't so fear inducing. And so there was sort of some human elements to it they brought, but also the science. Because they were very motivated, they were able to grasp really quite sophisticated concepts.[8]

The process took considerable resources and time, but the outcome was a major success in terms of communicating heath information effectively. The website increased transparency and understanding of the data for parents, journalists, and clinicians and was commended in the media and scientific press.[9] A parent who was invited to participate in the design of the website wrote: "When (my child) was diagnosed, it was a trying time and I was exhausted by the decision-making. . . . There had been confusing headlines and scares about certain units and there was no clear information on hospitals' surgical statistics and how the

[National Health Service] monitors units. . . . I also now understand what the existing data on children's heart surgery can and can't tell us."[10]

If a Scientist Does an Experiment in the Forest but No One Is There to See It. . .?

We all know the adage about the tree falling in the forest. If no one is there to hear it fall, does it make a sound? A similar philosophical question can be asked of what constitutes science—if it is not shared with others, is it science?

In the fall, I teach a graduate course on research methods. At the start of the semester, I give a brief lecture on the history of science and ask my students to reflect on what constitutes "science"—when and where did it begin? As part of this reflection, I lay a rudimentary timeline across the classroom floor. On one end is the present day, and at the other end is fifty-five million years ago, when primates began to evolve. Along the timeline are crucial developments, such as the discovery of fire, the emergence of art, the scientific revolution, and the creation of the printing press. I then ask students to stand along the place on the timeline they think most closely resonates with where they believe science began. Interestingly, the students tend to converge in one spot along the timeline.

That place is quite early in the history of human development—about two hundred thousand years ago—long before the word *science* first came into popular use in the nineteenth century.[11] What was so special about this particular place in time? It was when humans first began to develop language and communicate.[12] In the ensuing class discussion, the students would explain that, for them, it wasn't the invention of key discoveries that most determined what science was; rather, science only became science when it was shared between humans.

My students were picking up on a pillar of science—that a researcher must share their theories and findings with others for the work to be considered scientific. This prerequisite of communication is at the heart of the practice. For example, the scientific method as we know it today is in large part due to the massive libraries of the Islamic world in the ninth and tenth centuries and the interest of a diverse group of scholars in translating the works of forgotten Greek philosophers.[13] In the modern day, we have two main ways of communicating our research that are recognized to "count" as a contribution to science: by publishing peer-reviewed papers and by presenting at gatherings of other researchers in our field. We can share our work in other ways if we choose (via social media, films, presentations to community groups, and so on), but these methods are largely seen to be extras. Such communication methods might be appreciated among those outside academia, but they will not get us tenure, and, indeed, spending a lot of time on such "popular science communication" can be perceived by some as detrimental to one's career. For example, in their book on evidence-based policy-making, policy scholars Karen Bogenschneider and Thomas Corbett tell the story of how a scholar was almost denied tenure at Harvard due to his "extra-academic publications":

> This scholar already had established an impeccable national reputation as a poverty researcher, but he also wanted to inform public policy. Toward that end, he had published a widely read book on welfare reform that was accessible to broader audiences and would be very influential in the Clinton administration when it would come to power a few years later. . . . During the tenure discussion, the president of Harvard spoke negatively regarding the scholar's prospects . . .; his views were largely based on the tenure candidate's writing of a book for popular consumption. This was not the work of a serious scholar, all his other traditional research notwithstanding.[14]

This case may seem extreme, but I would wager that most academics who have attempted to stray beyond conventional scholarly methods of communication have had their knuckles rapped at least once or twice for doing so. For example, toward the end of my PhD studies, a couple of senior researchers attempted to dissuade me from returning to Bolivia to share the results of my research. They were concerned that I needed to focus on publishing papers and getting a job (and they weren't wrong), but it seemed incredible to me that—given that my research topic was about the potential local impacts of research—they would attempt to dissuade me from sharing back results!

To put the concerns and recommendations of these mentors in context, they were simply guiding me to follow the well-trodden path of what it takes to make it as a researcher. "Publish or perish" isn't just something people say, it's what defines success in academia in many parts of the world. Researchers who neglect to regularly publish in the peer-reviewed literature do so at their peril. A few years ago, I interviewed for a tenure-track position at a top research university in my region. While visiting the campus, it was made clear to me that my previous publishing record of two to three papers a year would not be enough to make tenure at the institution. Nothing was said about the quality of the work I was expected to do or its potential value for society—it was the quantity of papers (and grants) that mattered.

Some have critiqued that one consequence of this hyperfocus on quantity is that too many papers are being published, and many of them are of such poor quality that they will never be cited.[15] This surplus also adds to the burden of reviewers and editors who must review all these papers for publication. Reviewers are not remunerated for their labor, and completing peer reviews of articles can take many hours—sometimes days—of focused work. According to one estimate, scientists contributed more than one hundred million hours of their time completing reviews in 2020, equivalent to more than fifteen thousand years and more than $1.5 billion in unpaid labor.[16] To add insult to injury, while scientists are volunteering their time and energy to complete reviews, academic publishers are reaping billions of dollars a year in profits.[17]

To address these issues, some scholars have suggested that it's time to reconsider what it means to publish peer-reviewed work or even abolish the practice altogether.[18] For example, there have been some calls to reduce the number of papers a prospective job candidate can submit for consideration and even to limit researchers to publishing only one or two articles per year.[19] Other efforts seek to broaden engagement in the assessment of scholarly articles, both prior to and after publication, by, for example, disclosing author or reviewer identity, creating interactive "public" forums for the posting of articles for a more open review process, and allowing postpublication peer review.[20]

Although implementing such recommendations might help improve the overall quality of published scholarship and reduce the burden of peer review, these changes do not provide avenues for researchers to share findings of small projects that might not have groundbreaking findings, but that could be useful or relevant to communities or decision makers in a particular place. For example, research that finds the presence of waterborne pathogens in a given lake might not be a major contribution to the scholarly literature, but it could be important information to people who use the lake for swimming and fishing.

Deciding What, and Whose, Knowledge Counts

According to the *Merriam-Webster Collegiate Dictionary*, peer review is "a process by which something proposed (as for research or publication) is evaluated by a group of experts in the appropriate field." Typically, we think of peer review as being conducted by other scholars with expertise in our discipline, which means that the group of individuals included in this attestation of "truth" is limited to those also in the social and professional ranks of those putting forward the experiments or papers. But this system begs a couple of questions. First, what constitutes an "expert in an appropriate field"? And second, beyond scholarly articles, are there other products of scientific knowledge that should undergo peer review?

When reflecting on these questions, there are a few things to consider. Perhaps the first is that of scientific validity. Knowledge, whether scientifically derived or otherwise, is only useful to the extent which it is valid. But validity is not cut-and-dried—it comes in multiple forms. For example, those of us who do research are familiar with external validity, which refers to the extent to which a given scientific finding can be said to be broadly applicable. In other words, do the findings apply in other contexts and geographical locations? Can they be replicated? External validity must be demonstrated in terms of how a given research study (or the policy-based recommendations of a given study) does or doesn't apply in a specific context. For example, in chapter 4, farmer Matt Bomgardner spoke about the research-based techniques in a book on grazing techniques that did not work for his farm because of differences in climate. Thus the test of whether a given theory has strong external validity is to the extent it can be relevant across a broad range of real-world applications.[21]

What about internal validity? Internal validity refers to the degree to which the results of a given study show what they claim to show. That's important because in any given study, there can be many confounding variables that could skew the results. An example is political polling data, which seeks to measure voter intention but may sometimes just be a measurement of the preferences of the type of person who would be willing to spend time taking a political survey. Making sense of polling data often means making unilateral assumptions about who is being sampled and then extrapolating evidence from prior years in terms of who is likely to show up to vote. That's why political polling has narrow confidence intervals and why pollsters frequently get it wrong—and sometimes really wrong.[22] For example, polling of the 2016 referendum of the peace deal to end Colombia's fifty-two-year conflict with the Fuerzas Armadas Revolucionarias de Colombia–Ejército del Pueblo overwhelmingly predicted that voters would support the measure. But when the measure narrowly failed, journalists and social scientists pointed to a "spiral of silence" in which opposition perspectives were publicly shamed, so many opponents of the peace agreement kept their opinions to themselves until voting day.[23]

Robust internal and external validity are thus essential not only to ensure that scientific research is credible, but also to determine if and how its evidence can be applied to policy and practice contexts. As suggested in the examples above, this step is not just a matter of dispassionate evaluation by outside experts; both external and internal validity largely rely on knowledge that is situated in a specific place and time. In other words, people making decisions about policies need to know whether those policies will apply to their local context.[24] This stage is where incorporating viewpoints from people who have a stake in the outcomes of research, whether such people are patients, community members, or practitioners, has the potential to increase both the internal and external validity of research.

For example, in the 1980s, the education system in the United States was experiencing a reckoning. A report published in 1983, "A Nation at Risk," spelled out the state of underperforming education, pointing to plummeting student performance in math and writing and demonstrating that, compared to other economically advanced countries around the world, US students were largely falling behind.[25] The report was both a reproach and a call to arms and spurred two decades of bipartisan debate about reforms to federal education policy and funding, culminating in the signing of the No Child Left Behind Act in 2001 by then US President George W. Bush.

One of the stated goals of the act was to use the evidence of "what works" to determine which schools should be restructured and how funding and other resources should be allocated. As said by Bush at the time, such resources would be "directed at methods that work. Not feel-good methods. Not sound-good methods. But methods that actually work."[26] In other words, educational policy would not be based on the subjective perceptions of the teachers and administrators who ran the schools; rather, it would be based on a cold, hard evaluation of the (quantitative) evidence. School districts would be accountable for how their students performed on standardized tests. Failure to meet performance targets had high stakes: teachers and administrators were fired, students relocated, and sometimes even schools were closed.

The "what works" approach also determined which solutions should be implemented to fix failing schools; funding was intended to be used for the implementation of "scientifically based" interventions.[27] But although the reforms were largely aimed at benefiting the students most struggling and to close the then-called racial and socioeconomic "achievement gaps,"[28] the impact was not as intended. One of the main issues was that the law was designed to treat all schools the same and thus did not take into consideration context-specific differences among schools. For example, students were required to meet a certain reading and math level by a certain timeline. If one school had students three years behind the expected level and hard work by teachers and administrators brought the students up to only one year behind, the school was still considered underperforming.[29] As a result, the guidelines were often perceived as an "impossible standard"; the schools with the largest number of struggling students were faced with repeated sanctions for failure to meet requirements. Teachers in struggling schools were forced to work overtime to continuously revise curriculum, and overstretched school budgets were directed toward test-focused tutoring. As a result, the act increasingly became known by educators as "No Teacher Left Standing." As described by Vivian Tseng, president of the Foundation for Child Development: "It was so frustrating to see that this evidence-based policy movement was doing the reverse of what was helpful. And, in fact, it was making folks on the ground feel very resentful of the evidence-based movement, because a punishing school accountability system was wrapped around test scores. And a lot of teachers, rightly or wrongly, felt like this 'scientifically based' movement was something that was being done to them, not with them."[30]

Ultimately, although the law succeeded in identifying failing schools, it was much less successful in fixing them. Interventions that worked in one place often did not work in another, largely because different schools faced different realities. Student populations had diverse reasons for falling behind.[31] In other words, the data could show which schools were failing, but it did not explain why that was and what specifically to do to improve them.[32] For example, one of the evidence-based solutions for fixing failing schools under No Child Left Behind was to offer free tutoring to students who were falling behind. But studies of this practice found that its effectiveness was extremely varied.[33] Although some schools found that tutoring improved student test scores, others found it to have no effect.[34] Thus, rather than blanket solutions to the seemingly similar problems that many schools seemed to be facing, it was increasingly clear that theories of what worked needed to take local environments and contexts into account.[35]

In response to the failures of No Child Left Behind, Tseng and other educational scholars argued that what was needed was to "democratize evidence" so that those on the ground—namely teachers, parents, and administrators—have a voice in determining not only which policies would work for their schools, but also in determining what evidence is needed and how to use it to improve their schools. Said Tseng: "Democratizing evidence means that the outcomes of the research, the knowledge, should not just be for other researchers. They should be for the public. And so that goes beyond open science, [because] making research available doesn't mean it's accessible or that people find it useful. And so what are the supports and tools that folks need to actually make use of research and data? How do you equip the public to be able to wield evidence so that it can be a source of power?"[36]

From "Evidence Bases" to "Evidence Uses"

Equipping the public not only to access evidence, but to use it, is a tall order for the research community, which has largely been steered toward the production of evidence but often not to thinking about its use. Even the so-called evidence-based policy movement has mainly focused on creating the infrastructure needed to build an evidence base (such as through synthesis reports and online databases) and less about what happens to evidence in the process of making decisions.[37] This path, though, has largely led us to the creation of a massive amount of information that has potential for impact but that often goes unused. For example, members of the Advisory Committee on Climate Change and Natural Resource Science determined that the best way to produce "actionable science" was to start by grounding research approaches in terms of what specific decisions needed to be made.[38] They made this determination after generating hundreds of vulnerability assessments, most of which had been requested by environmental managers. When they delivered the assessments, however, many of the managers were unsure of what to do with them. It became clear that for science to produce actionable results, both decision makers and scientists needed to work together to think through which "products" might be most useful. As members of the advisory committee described: "Managers can explain the decision or planning issue at hand, the legal, political, social, and fiscal constraints, and explain how scientific information affects their decisions and downstream decisions. Scientists can ensure that the right product is developed and that managers understand how to appropriately use the information. Stakeholders (industry, landowners, potential downstream users of the information, and other persons who might be affected by the decisions) can provide insights on practical constraints and alternative courses of action that might affect the decisions and the science needed."[39]

Thus the call to democratize evidence suggests a reevaluation of not only who should be counted as an "expert in an appropriate field," but of what constitutes a scientific "product" that is worthy of peer review. This latter point is also important because what matters most from a particular study is often a question of perspective.

For example, during the last couple of years of my PhD studies, I'd become increasingly interested in budget filmmaking as a way of sharing research results, so I created two short films based on my research at that time. One film showed brief interviews with Indigenous leaders about their perceptions of research, and another depicted the work and challenges of being a park guard in Bolivia.[40] Both films were well-received, and the park guards loved the latter video and showed it when visiting communities around the park. Based on such positive feedback, I then decided to put together a film that would be an overview of my experiences as a researcher in the communities. I called it "Yo Soy Investigadora" (I Am a Researcher) and used a point-of-view approach to emphasize that I was attempting to show what I learned from my perspective. In other words, I strapped the camera to my head and walked around, trying to put a spotlight on the "researcher gaze."

The idea was somewhat sound. One of the problems I had increasingly encountered during my time in Bolivia was that many people didn't understand what research was or why people did research. For example, although people from communities in the region told me that they had problems with too many researchers, they also frequently asked me to explain to them in clear terms what research actually meant and what it was for.[41] I thought that showing "A Day in the Life" and explaining why and how I was doing research for my PhD could help increase understanding for people outside of formal research settings to aid people in the Madidi region in negotiating rights and responsibilities with different types of social and natural scientists.

However, although the idea made sense to myself and other foreign and Bolivian researchers with whom I shared early versions of the film, people in the rural communities for whom I'd ostensibly produced the film didn't seem to know what to make of it. Alas, I think I more likely showed people what the inside of a burnt-out PhD student's brain looked like. Aside from a short three-minute segment in which I show clips of interviews with Indigenous leaders and park managers about their perceptions of research, the film is the ultimate example of academic navel-gazing. I wake up in a comfortable house in the British countryside and then ride my bike along a canal to a well-funded university where I enter an office full of papers. I greet friends in a soil lab, and on and on. Friends of mine, who knew I had my heart in the right place (even if my head was somewhere else), asked gentle questions like, "But where's the rest?"

Ugh. What had I been thinking?!

In retrospect, I realize I had been thinking about what I thought would be useful from my own perspective rather than asking those who were involved in the research in Bolivia what would be useful to them. One problem researchers have is that the meaning of research is not the same to all those involved; as discussed in chapter 3, what constitutes a "useful" outcome may differ based on one's position in relation to the research.[42] This can create a real tension between the types of products thought up and produced by the researcher compared to those sought out by those who participated in, but did not lead, the research. One challenge for researchers, particularly students and early-career researchers, is that there is little guidance available in terms of how to create research "products" other than the standard scholarly publications, conference slide decks, and scientific posters. Due to lack of experience, senior researchers and mentors may not have the expertise to assist their advisees and mentees in the inevitable challenges and tensions that may arise when disseminating the results of research in extra academic settings. This problem points to a need for additional training and support from organizations that do have such expertise. For example, in the opening story in the chapter, some of the lead researchers had very limited experience in doing public communication, but their funding source enabled them the opportunity to work with the organization Sense about Science, which had a proven track record of creating opportunities for publics to engage with science.

Funding and Incentives

The examples above highlight the importance of engaging the end users of research in its evaluation and application, but there are two big caveats to this: funding and incentives. As my PhD experience shows, researchers are generally not incentivized, and are often discouraged, from sharing their science in nontraditional formats. Perhaps more vitally, even those who go against the grain may struggle to find funding sources to support their efforts. Such was the case for me; I was unsuccessful in obtaining any funds to pay for dissemination efforts and thus largely paid for them out of my own savings. Perhaps the case could be made that I was a junior researcher at the time and had not yet proven myself enough to obtain this type of funding, but even for senior researchers, grant support for "extra-academic" research communication is very limited. For example, Pagel told me that since first developing the Children's Heart Surgery Outcomes website, she has sought additional funding to expand it to include additional information that would be useful to parents, but has thus far been unsuccessful: "No one's willing to fund it because it's not primary research and it's not public engagement in the way a lot of funders see public engagement, which to them is scientists holding a public event and talking at people. And so basically everyone's saying that's not my responsibility, and what that means is that it's not getting done. And I think that's a real shame."[43]

Some initiatives to change are under way, however. For example, the Transforming Evidence Funders Network brings together public and private funders across the world and across disciplines to share funding practices focused on improving the use of evidence in policy and practice.[44] Funders in the network learn from one another about how to support policy-relevant science, often by playing a brokering role themselves in the building of relationships between decision makers and researchers. Although there is a lot of information available on *what* funders support, there is less guidance in terms of *how* funders can further the effective use of evidence in real world applications. Members of the Transforming Evidence Funders Network share best practices for supporting researchers engaged in this type of work, including by establishing a pathway for achieving both the nonscholarly and scholarly impacts of the research at the project outset and by providing key points to revisit these moments throughout the project.[45]

Related to funding are the incentives that drive academic production. As noted above, tenure and promotion within academia are largely determined by the peer-reviewed publishing and grant-awarding system; other types of academic "products" typically don't count. One effort to change this emphasis is the Community-Engaged Scholarship for Health website, an online portal overseen by an editorial board comprised of both researchers and practitioners from nonprofits organizations, community-based organizations, academia, and government.[46] The website seeks to be a resource for academics seeking to peer review their community-oriented (nonmanuscript) products of their research (such as curricula, policy briefs, and videos) and thus increase the visibility of these research products for promotion and tenure.[47]

Within universities, some tenure and promotion committees are also making changes by providing guidance on how to credit nontraditional products of community-engaged and policy-relevant scholarship. For example, a multidisciplinary team of faculty, staff, and administrators at the University of Louisville in Kentucky created a faculty handbook for the evaluation of community-engaged scholarship to help with the understanding and assessment of research that involves communities.[48] Funders can also play an important role in promoting such work within academia. Several major funders in the United States created the Institutional Challenge Grant as a way of encouraging university-based research institutions, schools, and centers to reward faculty partnerships with practitioners and policy makers.[49]

A final point is that much actionable science is now happening beyond universities, which do not often operate in a manner able to support iterative and dynamic research that require being able to test approaches out quickly and to move on as needed, sometimes adjusting budgets and timelines in the process. University administrative offices are not typically known for either efficiency or flexibility; they move like tankers in open sea—slow, steady, and hard to change course—which means that researchers who seek to have impact beyond academia must often move elsewhere. For example, Piyush Tantia was the founding chief executive officer and is now chief innovation officer of ideas42, a nonprofit organization that uses research from the behavioral sciences to address social problems, collaborating directly with governments, industry, and nonprofits.[50] But before ideas42 became a nonprofit organization, it was a project housed within a large research center at Harvard University. Tantia told me that the overwhelming bureaucracy, as well as financial considerations, made moving outside of academia a necessity.[51] He explained: "As an independent nonprofit, we are nimbler and can prioritize social impact while still maintaining academic rigor in our work. We are also able to partner with academics from several different universities around the world, which is harder as a university-based center. University-based models can work, however, when the university realizes the needs of field research and social innovation. Such field work also provides an opportunity for experiential learning—something that is in high demand from students."[52]

Tantia's latter point is particularly important. Universities have one great asset—people—and these people come in the form of students, faculty, and staff who are dedicated to and passionate about the mission of learning. Involving students in collaborative research with community groups and other local organizations can be a powerful way to show students that both the challenges and opportunities can emerge when research is designed to be relevant to the needs of community leaders and other decision makers.[53] Perhaps most importantly, training students to think differently about who should have a voice in determining the products and processes of research can provide the next generation of researchers and practitioners with foundational experience for how to navigate such partnerships in their own academic careers and beyond.

PART THREE

The End Is Just the Beginning

"The single biggest problem in communication is the illusion that it has taken place."
—George Bernard Shaw

"Listening is hard . . . because you run the risk of having to change the way you see the world."
—Zia Haider Rahman

CHAPTER 8

The Scientist Next Door: Conversations, Communities, and Connections

THE FIRST PERSON TO TEACH ME the value of conversations about science was David Burg. Burg was no academic, but he deeply valued the role of science in improving environmental management. He ran a never-quite-official nonprofit based in New York City called Wild-Metro, which used scientific evidence and nature-based excursions to engage people in caring for their local environments. But as much as Burg loved his yellow warblers and eastern meadow voles, he loved people, and he especially loved talking to them.

When we set off to remove invasive multiflora rose that was encroaching on the habitat of a 450-year-old "Granny oak" in his favorite park in the Bronx, it could take us an hour to walk the quarter mile from the parking lot to the site. Burg would stop to talk with anyone we passed—a mom pushing a stroller or a group of energetic teenagers. He would smile at them and say hello, and if they responded he would immediately ask them if they knew about the parakeets in the trees above or if they had gone to see the harbor seals off Hunter Island.

New York City isn't a place exactly known for striking up conversations with strangers, and initially these interactions made me a bit uncomfortable. I would squirm as people struggled to make sense of this older man, dressed in ill-fitting khakis and wearing binoculars around his neck, peppering them with questions about things many people might not often think about. But more often than not, they would begin to look up and around them—into the trees where Burg pointed or even on the grass where a robin pecked for worms. Sometimes they would even ask to look through his binoculars, and every so often, a stranger would be so interested that they'd accompany us the rest of the way to Burg's beloved oak tree. Indeed, the volunteer base for Burg's organization largely came from interactions he had in urban parks with strangers.

As I noticed the way that people responded to my friend—with curiosity, questions, and sometimes requests for more information—I realized that his method of engaging people was powerful. Not until many years later, however, did I realize that chatting up strangers about urban wildlife wasn't just a personality quirk, it was a sincere way of communicating about science.

The Community in Communication

Of all the things scientists are likely to hear about "fixing" the relationship between science and society, it is that scientists need to better communicate with the public. Google "scientists are bad communicators" and you will get pages upon pages of results, many in the form of opinion articles and blog posts that cite examples of a personal experience with a scientist who was so ineffective at explaining what their research was about that they ended up turning the other person off science altogether. Having spent much of my life in academia, I can relate to some of these stories and, indeed, perhaps have been guilty at times of being similarly obscure and pedantic when attempting to describe my research. Once, as a master's student, a woman asked me what my thesis was about, and after I rambled on for a bit about concepts like empowerment and sustainability, she said, "All I heard was words."

So yes, researchers are not always great at explaining what we do and why it is important. To help us get better at it, an increasing number of science communication training programs are available to help "translate" complicated scientific concepts into language that people unfamiliar with such ideas can understand. Such programs provide training for researchers about how to talk to journalists, present their science to policy makers, and write op-ed pieces for newspapers, among other skills. This work is necessary and important. However, there is much more to good science communication than disseminating scientific concepts to laypeople in simple, distilled formats.[1]

What is *communication*? The term is used in so many contexts and to describe so many different activities that it's hard to pin down a specific meaning. Indeed, dictionary definitions will not help us much here, where the word is broadly used to refer to various processes, systems, outputs, and connections. Neither have scholars in the fields of science communication been able to come to a consensus on an all-encompassing definition.[2] Increasingly, though, there is recognition that communication is not some one-and-done approach to sharing information (no matter how fancy the presentation); rather, it requires a two-way exchange of ideas.[3] Thus using the term *communication* to describe activities that involve a one-way transfer of information, with little or no opportunity for dialogue, could be seen as a misuse of the word.[4] Imagine trying to "communicate" with your romantic partner through a podcast in which you explain the various ways that their lack of punctuality impacts your life or "communicating" with your teenager about doing better in school through an inspirational documentary. Effective communication with our loved ones requires not only telling them what's on our mind, but listening to what they have to say as well. Why should science communication be any different?

An effective way to think about what we mean by science communication is in terms of what we're hoping to achieve.[5] Many scientists wish to be skillful at communicating with the public because they want to persuade people about the "facts" of science, particularly if research suggests some important societal changes that should be made. This goal is valid, but it highlights a major mismatch in terms of what most researchers are doing in science communication versus what has been shown to work in convincing people to change their attitudes and behaviors. As explained in chapter 2, one crucial ingredient in shifting minds, social norms, and related behaviors is not facts, but social confirmation. Thus researchers and others who wish to communicate scientific information can benefit from understanding how social confirmation shapes how people interpret scientific ideas and how shifts in people's social networks can enable them to see "facts" in a different light.

One story that illustrates the importance of social confirmation for influencing perceptions was shared with me by Emily Brown, an artist and counselor based in the southern part of the United States.[6] Brown grew up and was very involved in conservative Evangelical Christianity and then, as a young mother, moved into a more liberal, "hippie" community of friends, where she homeschooled her children. She became a mother in a time and place when more "natural" ways of childbirth and rearing, such as the use of midwives and breastfeeding, were largely dismissed by the medical community in her region, and she remembered how women were arrested for breastfeeding in public spaces.[7] Her community at that time was largely made up of other mothers who were skeptical of the medical establishments' promotion of cesarean sections over more natural forms of childbirth, and this developed into other forms of skepticism, including hesitancy toward vaccinating their children. She met regularly with her group of trusted mothers, and together they searched for answers to the questions that concerned them about vaccines. As she described it: "I had been accepted into a group of strong women who didn't vaccinate. We were doing a thing together, and we had really believed in what we were doing. We were on a mission. We were fighting. We founded an organization to support midwifery and natural birth. We were trying to make a change."[8]

Brown said that while being part of this community, her beliefs about vaccination, among other things, began to shift. One of the most powerful moments was when a couple of other mothers in the group mentioned watching a television program about the history of the polio vaccine and said that it had caused them to question whether they were doing the right thing in not vaccinating their children. As Brown didn't have a television at the time, she wasn't able to watch the program, but just hearing others in her community voice their doubts made her begin to question her own beliefs. The biggest change came when she enrolled her children in public school, where standard vaccinations were required. Although she could have claimed a religious exemption to avoid vaccinating her children, she felt uncomfortable doing so, as her hesitancy was not religiously motivated. At the same time, Brown was experiencing other personal changes in her life. She had recently moved to another state and joined a different church community. She said: "My kids grew up, and everybody started going to school. I got more involved in my art and less involved in homeschooling. I had become an Episcopalian. Over time, things shifted, and I had different people around me. I still had that [skeptical] element there. But it was less."[9]

Brown ultimately decided to vaccinate her children, and she now sees herself largely on the opposite side of the vaccine debate. She has watched as her more Christian, conservative acquaintances have taken up the antivaccine stance previously touted by her secular back-to-nature friends. She said that she felt extremely frustrated during the COVID-19 pandemic to see so many people in her part of the country refusing to get the vaccine, some who later died from the disease. COVID-19 skepticism had become an important aspect of being part of the Evangelical community, and she knew of people who were sick but who hid that from friends and neighbors. In one case, an old friend of hers suffered debilitating long COVID symptoms but became hesitant to speak openly about it because of backlash from her community.

A few things struck me about Brown's story. First, the skepticism of vaccines that she and her friends shared wasn't absolute. She said that the group of mothers she belonged to was regularly seeking out information to help them understand what was best for their children and discussing it with one another. Brown also told me that she believed she would have vaccinated her children much sooner if she could have gotten a doctor or scientist to patiently listen to her and answer her vaccine-related questions. She explained that when she had tried to bring up questions with members of the medical establishment, she had been largely dismissed. Brown said: "I felt like if they would just have been patient with me and been like, okay, this is confusing. Let's go through it step by step. . . . If they had really engaged, not in a combative way, and been like, 'Well, that's one viewpoint. Let me show the other viewpoint' . . . that probably would've been more effective with me."[10] Brown's experience points to the importance of seeing science communication as an opportunity for listening and dialogue rather than just a clear explanation of the facts. Finally, her story demonstrates the importance of community and connection in shaping beliefs. As her social networks changed, so did her ideas about vaccination. Her questioning was supported and validated.

In *How Minds Change: The New Science of Belief, Opinion and Persuasion* (mentioned in chapter 1), author David McRaney makes the case that changing one's mind about something, particularly if that something is an emotionally charged or controversial issue, requires a lot more than an influx of counterarguments.[11] Rather, he suggests that people can begin to change their own minds when they are given the opportunity to connect on a human level with someone who espouses the belief that they largely reject. Such encounters, if thoughtfully managed, can help people engage more deeply with their own processes of inquiry, opening the door to new ways of thinking. In other words, creating opportunities for people from different communities to listen to one another is an essential component of communication that is both impactful and persuasive.

So what do Brown's story and McRaney's conclusions about the importance of direct, interpersonal connections for changing minds have to offer for thinking about science communication? Is there a broader lesson about how, and where, we are having conservations with the public about science?

The Man in the Lab Coat

Since the mid-twentieth century, researchers have been interested in the public's perception of scientists. In 1961, a study found that most undergraduate students held to the stereotype of scientists as "unsociable, introverted, and possessing few, if any, friends."[12] He (as the typical pronoun used at the time for such a profession was masculine) was highly intelligent, but also potentially prone to impulsivity and radical politics. Furthermore, he was undatable—as one student suggested, "maybe it's not a good idea for him to be married."[13] Related studies had children draw pictures of scientists, which led to decades of drawings of middle-aged White men in lab coats, doing experiments alone in labs. These patterns have held up across different cultural contexts: studies in China, India, Australia found similar impressions and images, and such stereotypes persisted into the twenty-first century.[14]

Although perceptions have begun (albeit slowly) to shift away from the unattractive man in the lab coat to include more women and people of color, perceptions of scientists being competent but not warm persist to the present day.[15] Part of these perceptions come from popular culture. In films, scientists are often depicted as removed from the world around them. They spend countless hours in labs, often alone, with one or two loved ones who understand them and give them their space—and sometimes a good talking to about being a full human being when they need it. But the key to their success is in being left alone to do the important thinking—in the isolation of their brains—that allows their genius to emerge. Only then, after Dr. So-and-So is pulling an all-nighter in the lab, scribbling on bits of paper until that eureka moment!—the world is saved. I've seen this fantasy depicted so often in movies about science that one of these days, just for fun, I'm going to run through the halls of my own university yelling "Eureka!" and throwing papers left and right.[16]

In such portrayals, scientists are not seen to be rubbing elbows with the commoner but are looking down from their ivory towers. Perhaps that's why a 2021 survey conducted in the United States found that 72 percent of respondents could not name a single living scientist.[17] Of the minority able to come up with the name of a living scientist, the most common answers refer more to experts working in public outreach roles, such as Bill Nye or Anthony Fauci, rather than active researchers. Similar results are found in terms of public awareness of the places where science is done; only about a third of respondents were able to name a specific scientific institution.[18]

For those of us who are part of scientific communities, such findings should give us serious pause. As an academic, I can name dozens of scientists, largely because they are my friends—they are part of my community, and I am part of theirs. My social network is full of scientists, and I bet that's true for most of us in academia. It's an incredible privilege to be part of such a community, in part because if I have a question about how the world works, I often have someone I can call to ask. When COVID-19 hit, I reached out to a colleague who is an epidemiologist; similarly, when I've had questions about math, physics, biology, or sociology, I have a handful of trusted friends who would be happy to share their knowledge.

But what does it mean that so few people can even name one living scientist, let alone have them listed among their cell phone contacts? I think it points to a major discrepancy between the communities that contain scientists and those that do not. In other words, while some communities are "rich" in access to science and scientists, others are more like "science deserts," devoid of opportunities to connect with researchers and participate in science-related activities.[19] For example, rural communities are particularly excluded from science. Studies have found that students in rural areas are substantially less likely to pursue science careers than their urban counterparts, or even to go to college.[20] Rural communities may have less access to both formal and informal science education opportunities, and less populated areas offer fewer job opportunities in science and technology, thus limiting exposure to "the diverse ways in which STEM is practiced in the world."[21]

Even in urban areas, those who are able (and who feel comfortable) to access spaces of science learning and communication tend to be from more affluent and ethnically or racially dominant backgrounds rather than from working-class and minority ethnic or racial backgrounds.[22] For example, science education and communication scholar Emily Dawson conducted extensive ethnographic research about perceptions of and participation in informal science education (such as science museums) with participants from Afro-Caribbean, Somali, Asian, Latin American, and Sierra Leonean communities in the United Kingdom. Her research found that an important—and often neglected—factor in understanding why people from underrepresented groups do not seek to participate in science learning is the strong sense of alienation that individuals from these groups may experience in such spaces. As expressed by one of Dawson's research participants to explain why he would not choose to go to visit a science museum, "You cannot connect with it. It's for those people that it matters to . . . the museum, or science itself."[23]

Perhaps unsurprisingly, research has found that distrust and skepticism toward science tend to be higher among individuals who report greater "psychological distance to science," which refers to the extent to which people believe that science is removed from or connected to their lives.[24] Social psychologists from the University of Amsterdam and Durham University surveyed more than sixteen hundred people in the United States and the United Kingdom to measure the links between one's "psychological distance to science" and science skepticism, asking respondents to respond to survey items such as "I rarely interact with scientists in real life," "Science and scientific research play a big role in my local area," and "I can see the effects of science, whether positive or negative, on the world."[25] The researchers found that "psychological distance" was a stronger predictor of science skepticism than political ideology, religion, and knowledge of science and that it was also a reliable predictor of one's science-related behaviors (such as whether to get the COVID-19 vaccine). These findings suggest that increasing one's "proximity to science" (which can be measured by asking questions such as whether someone personally knows a scientist) can reduce skepticism and contribute to more positive attitudes toward science.[26]

A Right to Science

Thus perhaps one goal of science communication should be to create more opportunities for people to engage with and learn about science in places where there are fewer scientists and institutions.[27] One way to increase such proximity to science is using mobile science installations, such as laboratories that are transported by or housed within a bus or a van. The history of mobile laboratories dates to 1906, when George Washington Carver used a Jesup agricultural wagon and field demonstrations to share knowledge about soil science with farmers. Carver, born into slavery, and despite being refused entry to multiple academic institutions because of his race, became a renowned agricultural scientist during his lifetime. He is also credited with promoting "extension" work across the southern United States and was especially dedicated to reaching farmers who had the least resources and access to new techniques.[28] As he said, "My idea is to help the 'man farthest down,' this is why I have made every process just as simply as I could to put it within his reach."[29]

Modern-day examples of such laboratories often have a similar aim of reaching communities that have been historically underserved by science. For example, the BioBus initiative uses buses to bring advanced lab-based research in neuroscience and environmental health to thousands of students, primarily in the United States but also in schools in Africa and the Middle East.[30] Another example comes from Joey Rodman, a science educator in Oklahoma, one of the poorest and most rural states in the United States.[31] For more than a decade, Rodman worked in museums, camps, and planetariums located in the urban and suburban parts of the state. But they began to realize that there was a large group of kids who weren't participating in such science programs—namely, those from the rural areas who had the least access to science. So Rodman built a portable planetarium and began touring around rural Oklahoma, usually visiting schools but sometimes also giving shows at public spaces, such as libraries and churches. These types of initiatives point to the importance of seeing opportunities for people to have conversations about science as a fundamental human right.[32]

Culturally relevant approaches to science communication can be especially powerful. For example, Nabanita Borah is an atmospheric scientist from Assam, India. Borah's doctoral research was focused on predicting rainfall during the monsoon season, information she believed could potentially be useful to farmers. But even though her research resulted in several publications, she became increasingly aware that the information she was producing would not be accessible or intelligible to rural communities—for example, in her home state of Assam. She believed strongly that such communities had a right to such information and decided to explore the use of locally relevant methods of storytelling to communicate her research. One approach she used is called Ojapali, a traditional folk dance believed to be one of the oldest art forms in Assam. Ojapali is performed in a group and combines songs, dancing, and acting; it is used for both entertainment and as an educational tool for conveying moral and spiritual lessons.[33]

To test the use of Ojapali as a vehicle for communicating about climate change, Borah worked with high school students and Ojapali performers in the rural village of Amrikhowa, Assam, which she documented in a short film about the process.[34] One of the most interesting parts of the film is her work with the Ojapali performers, who initially were very skeptical about using the art form to communicate about climate change, a largely unfamiliar concept to them, but by the end of the process, they had changed their minds about the experience. Perhaps more importantly, they gave critical feedback on the aspects of the experience that they believed could be improved for future performances. Borah's work thus points to not only the value of developing culturally relevant storytelling methods for communicating science, but also the need for iteration and feedback in the creation process, which requires sustained engagement and support over the long term.

Engaging Skeptics

Researchers from various disciplines are beginning to take notice of two-way communication approaches and the value of interpersonal conversations for changing people's minds about scientific issues. In his book *How to Talk to a Science Denier*, science philosopher Lee McIntyre describes how he took the relatively extreme step of going to a flat earth convention to try to build personal relationships with attendees.[35] Similarly, climate scientist Scott Denning agreed to be a speaker at the Heartland Institute conference in 2011, despite receiving emails from colleagues discouraging him from, what seemed to them, an endorsement of climate change denial. Of his decision to attend, he writes: "Refusing to engage dismissive voices on climate change may feel like taking the high road, but I suspect it's the high road to ruin. Ignoring climate contrarians has not made them go away. In fact, their message has resonated with an increasing slice of public opinion for several years. . . . It seems to me that strong and persuasive engagement of that audience by more bona fide experts articulating the scientific consensus is essential."[36]

This engagement is particularly important for debunking misinformation. Scholars who study the best ways to correct misinformation often distinguish between "topic rebuttal" and "technique rebuttal." Topic rebuttal is something all scientists are familiar with; it means contradicting erroneous information with facts based in evidence and scientific consensus and thus requires knowing the science extremely well so that one is prepared for challenges to the contrary. Technique rebuttal is entirely different. Rather than articulating the evidence, it focuses on getting the other person to examine their thought process, which can uncover logical fallacies in decision-making. Dialogue methods that utilize a form of technique rebuttal can inspire the participant to reflect not only on what they believe, but how they came to believe it. Research has found that both topic and technique rebuttal can be effective for debunking misinformation, which can spread if not addressed.[37]

However, it is crucial that refuting misinformation be done in a respectful and kind manner; otherwise, it will simply turn people off. McIntyre realized that the hard way when he first confronted a speaker at the 2018 Flat Earth International Conference and attempted to poke holes in his reasoning. When the speaker turned away to talk with someone else, McIntyre was made aware of the flaw in his approach. He explained, "Looking back, maybe I don't blame him. I was too hot. Too confrontational. It's hard to stay cool when your beliefs are being challenged. Maybe I was proof of that myself."[38]

A final approach that could be useful for science communication is called "deep canvassing," a method whereby people who advocate for a certain position on a given issue go door to door to have one-on-one conversations with members of the public.[39] The concept and practice of deep canvassing was first developed by the LGBTQ+ community in Los Angeles to try to understand voters' concerns about same-sex marriage. Unlike traditional canvassing, which seeks out people who already are in support of one's position and encourages them to engage further or which tries to use facts and information to sway existing opinions, deep canvassing does something different. Here, canvassers go to communities where they expect to find a stronghold of opinions that differ from their own. When they knock on doors, they are trained to ask questions to better understand the experiences of the other person in relation to the issue. They engage in empathetic listening and share their personal stories about the issue in question.

Although such conversations only last, on average, about ten minutes, they can be deeply impactful. In 2016, researchers found that a deep canvassing campaign designed to increase empathy toward marginalized groups, such as unauthorized immigrants or transgender individuals, reduced prejudicial attitudes among participants, an effect that lasted at least several months after the conversation. In contrast, traditional door-to-door approaches had no measurable impact.[40] The researchers found that a crucial component was the nonjudgmental exchange of narratives; when this part of the process was removed, the effectiveness of the intervention was significantly reduced. In other words, the canvassers had to be able to listen with an open mind and to share their own personal stories for the approach to lead to a mental shift.

Neighbours United, an environmentally focused nonprofit organization based in Canada, developed a similar approach to engage people from rural communities in British Columbia in conversations about climate change.[41] Staff from the organization decided to pilot the method in a community called Trail, which is home to one of the largest zinc and lead smelters in the world and the largest employer in town. Prior efforts to engage community members had not been effective, as taking action on climate was largely seen as being in conflict with how many people made their living. Despite this challenging environment, the canvassers set an ambitious goal to get the city council to vote in favor of transitioning the town to 100 percent renewable energy for all community uses.[42] They developed a script that included opportunities to listen to the concerns of neighbors, for canvassers to share their own stories of climate change already impacting their lives, and to draw out the lived experiences of the person at the door in how climate change was already affecting them or their loved ones. In addition, because building a sense of community and overcoming the common feeling that tackling pollution problems like climate change were too overwhelming, canvassers reminisced about a 1980s environmental success story in Trail when the community came together with industry and governments to reduce levels of lead in their air, land, and water.[43] To close the conversation, the canvasser asked if the community member would agree to support the clean energy resolution via a petition.

The results were striking. At the doors, organizers found that more than one in three residents (40.6 percent) were changing their minds about the importance of taking action on climate change.[44] In 2022, Trail's city council voted unanimously to transition to 100 percent renewable energy by 2050 and instructed staff to develop a transition plan to do so. As described by the project leader, Montana Burgess: "It changed the community narrative. Talking about climate change in Trail today is very different than it was four years ago. It's a much easier conversation."[45]

Being Intentional

Knocking on strangers' doors or attending science contrarian conferences is not everybody's cup of tea. Such activities perhaps belong more in the realm of science educator work than tasks to be undertaken by active researchers. But I mention these approaches because they offer a few important lessons for science communication more broadly.

First, it is important to be intentional by what one means when talking about communication. Science communication can have many different aims; researchers may wish to raise awareness about a certain issue, debunk misinformation, or change science-related behaviors.[46] In particular, the examples in this chapter suggest that it is important to be clear on which "publics" one seeks to engage.[47] For example, if one seeks to provide the latest research to an already science-minded public audience, speaking with a reporter from the *New York Times* or giving a public talk at a natural history museum would be a good approach. However, if the aim is to reach members of the public who may not have much access to science, it is essential to find other venues and approaches. Furthermore, it is not just a matter of being intentional about *who* to communicate with, but being very clear on *why* one is doing so. John Besley, science communication scholar and author of the book *Strategic Science Communication*, argues that such intentionality is essential for effective communication. "For example," he said, "a scientist might realize that the best way to communicate is not through a radio campaign, but perhaps by putting flyers on the backs of toilet stalls!"[48]

Second, one thing we have a bit wrong in science communication is the idea that "if we build it, they will come." Certainly, science museums, universities, and other informal science education centers can provide opportunities for people to learn about and interact with science. But while millions of dollars go into infrastructure—often in urban hubs—what is frequently neglected is funding for science programming, particularly programing that extends into places where there are comparatively fewer conversations about science happening. Although the types of activities mentioned in this chapter do not require multimillion-dollar educational centers, they do require investment in people and training.

For example, the deep canvassing approach used by Neighbours United required sixty-one iterations of the script, eighty volunteers, and eight hundred hours in the field.[49] Thus communication should not be seen as an add-on but rather an activity deserving of time, training, and funding in its own right.

Finally, perhaps the most important lesson is, quite simply, that interpersonal conversations about science matter. They matter for laypeople, and they matter for scientists who engage in these spaces. Two-way dialogue means that impacts go in both directions. Listening to the stories and experiences of others can teach us that people have much more nuanced views than appear on the surface.[50] They can also challenge us to rethink our own practices and ideas. Perhaps most importantly, they can communicate a feeling that science is inclusive and open.

Recently I heard a story from Julia Minson, a decision scientist at the Harvard Kennedy School of Government, that illustrates the importance of this idea.[51] When she was a graduate student, her mother was diagnosed with late-stage breast cancer. Desperate for solutions, Minson searched online and found information about an experimental new drug that offered a small chance of success. When Minson mentioned the drug to her mother's oncologist, the doctor was skeptical and told her that it was too risky to consider. However, a couple of weeks later at her mother's next appointment, the doctor said that she had brought the suggestion to the "tumor board"—a gathering of top oncologists in the region—who agreed with the doctor's assessment that the drug was too dangerous to try. Although disappointed with the outcome, Minson was struck that the doctor had listened and considered her perspective:

> I still remember that conversation—17 years later—as the time where I felt most heard, perhaps in my life. I didn't get what I wanted; she didn't change her mind. But she sort of put her career on the line; she pitched this crazy idea by a second-year grad student. . . . My entire

research program right now is receptiveness to opposing views, and I think part of the reason that story is really precious to me is because I spend a lot of time trying to convince people that making somebody feel heard doesn't [necessarily] require changing your mind.[52]

What this story—and many of the examples in this chapter—suggests is that listening to others' perceptions of the issue at hand, even if we disagree with them, can be a powerful form of science communication.[53] In the words of science communicator Faith Kearns, doing so can mean "centering the role of relationships in science communication, not viewing them as peripheral or separate."[54] There are endless physical and virtual spaces where we can simultaneously wear our "scientist" hats along with any others that seem relevant: parent, neighbor, school board member, community activist, parishioner, teammate, coworker. As my friend David Burg taught me all those years ago, you don't need to have to be standing on a stage to talk about science. Sometimes, the most important spaces for doing so are right in our own backyard.

CHAPTER 9

The Skeptic in the Mirror: The Essential Role of Uncertainty in Science

I REMEMBER, AS I'M SURE MANY OF US DO, the moment I first realized COVID-19 was serious, perhaps more serious than most of us could imagine. It was March 2020, the week before spring break, and I was in a small classroom in New York City with a group of undergraduate students. The news about COVID-19 was constant yet unclear. As faculty, we were told to prepare for campus closures, but we were also sent notices that we should largely carry on as usual until indicated otherwise. I'm a stickler for class attendance and participation, as my students well know, so it was in front of a very full classroom that I found myself with my students looking up at me and asking me what I thought would happen.

I had no idea what to tell them. I know as much about public health matters as your average person. Prior to COVID-19, I had never quarantined due to illness for more than a couple of days, much less lived through a global pandemic. I vaguely remembered past scares about so-called superdiseases such as Ebola and bird flu and how they had been exaggerated by the media and then faded away just as quickly. The idea that we would all have to lock down in our homes seemed extreme and, frankly, very impractical. After all, we lived in New York City, not in some remote hamlet in the mountains. Where were we all supposed to go?

So when my students asked me for my opinion, I said, foolishly and yet with some degree of unwarranted confidence, "I have a feeling that this will all just blow over." They seemed momentarily relieved. We had a nice class, wished each other a good spring break, and about an hour later received an email from the university president announcing that the campuses were closing and that all classes would switch to remote learning. It was another year and a half, and millions of deaths later, before classes fully resumed in person.

In retrospect, it was both unprofessional and reckless for me to tell my students, from my position of relative authority, that I thought everything would be fine, yet I remember feeling like I had to tell them something. I was their professor and therefore someone who should "know things." Perhaps I had access to more information than they did. Maybe I should have known more. But I didn't know anything, and, at the time, pretty much no one did. What I should have said was that I simply didn't know, that we were all in the same boat of not knowing what was going to happen, and that the best thing to do would be to take the precautions being recommended by health experts. But no one likes to say they don't know. Just as humans are hardwired to "see" facts through the lens of our preexisting beliefs and biases, we are inclined to seek answers—even unsatisfactory answers—to resolve doubt.[1] My students regularly sought my advice, and I regularly gave it to them; that felt like part of the contract of what being a college professor meant. In hindsight, I realize that not only was I giving them factually wrong information, I was also doing them a larger disservice in relation to science communication. Instead of being honest and up front about what I didn't know, I was making firm statements based on unfounded assumptions from a position of authority. However, I was not alone in my hubris.

The lessons of COVID-19 are not just for epidemiologists and public health experts. Reflecting on our attitudes and behaviors during that time—what we thought we knew, what we got wrong, and how we interacted with others—offer us all an opportunity to think about how we talk about (and perhaps think about) scientific knowledge and how we speak to others from our positions within the academic community. Perhaps there is no better example than in terms of how we thought about one particular aspect of the pandemic: face masks.

To Mask or Not to Mask

The medical face mask—a thin piece of fabric designed to limit the airborne transmission of pathogens between wearers—has become an unprecedented symbol of various things—some science related and some not—across the world.[2] In places where masks were mandated to reduce the spread of COVID-19, protests broke out, notably with signs such as "Masks: The New Symbol of Tyranny" and "Free Your Face!" In some countries, mask wearing became an expression of one's political identity and thus, one's so-called tribe. For example, in progressive communities in the United States, masks became a sign of taking the pandemic seriously and of sacrificing one's personal liberty to help save lives. Conversely, in conservative circles, the mask became a symbol of government overreach, contradictory policies, and infringement on one's first amendment rights.[3]

Things got heated. In Australia, one man reportedly suffered a heart attack when police put him in handcuffs for not wearing a mask while exercising outdoors.[4] In California, a parent who was angry about mask requirements for small children punched his daughter's teacher in the face.[5] And in Germany, a gas station cashier was shot in the head and killed after asking a man to comply with the rules and wear a mask.[6]

Fueling the debate was confusing, and sometimes contradictory, information coming from the medical research community. Early in the pandemic, the world's leading health bodies, including the US Centers for Disease Control and Prevention and the World Health Organization, did not recommend mask wearing for the general population due to a shortage of personal protective equipment and because initially the disease was largely thought to spread via surfaces. As the airborne nature of transmission became better understood, masks were recommended and then mandated in many places. Then, as vaccines were rolled out, the advice changed again, and vaccinated individuals were told they did not need to mask because vaccines were initially thought to prevent transmission. When that was found to not be the case, masks were again required.[7]

This flip-flopping back and forth made sense in terms of the relevant health bodies relying on the latest findings, but it was incredibly confusing to the public, who just wanted a straight answer—do masks work or not? Then a report emerged in early 2023 that seemed to settle the debate. The report was put out by the Cochrane Institution, which conducts highly respected metanalyses of scientific topics by aggregating the findings of randomized trials to provide a synthesis of what the research is saying and is considered the "gold standard" for meta-analytic reviews. The report described a review of seventy-eight randomized control trials and concluded that wearing masks "probably makes little or no difference" for preventing the spread of viral respiratory illness such as COVID-19.[8]

Given the political firestorm around masks raging around the world, the report was widely shared in the media as definitive "proof" that masks don't work. This interpretation was shared by the lead author of the review, Tom Jefferson. In an interview that came out around the same time as the review, Jefferson said, "There is just no evidence that they make any difference. Full stop."[9]

So were the antimaskers right? We'll come back to this point later in the chapter. But the most interesting thing about the Cochrane report wasn't about masks and wasn't even about whether the conclusions were valid. Rather, it was about something that is perhaps at the heart of research and perhaps the main thing that distinguishes the scientific method from religious faith. It is not hypotheses, observations, or even treatments versus controls. Instead, it is something we often don't like to talk about much in science: uncertainty.

Science Proves Nothing

One of the biggest misconceptions about science is the idea that it proves something. We see this "proves" wording constantly in the news, even in otherwise reputable popular science magazines such as *Scientific American* and *National Geographic*.[10] "This study proves that . . ." reads the headline or the concluding paragraph. We hear it on the radio and, yes, even on NPR and BBC. Perhaps even more disconcertingly, scientists themselves have been known to say that their research "proves" this or that; even high-impact science journals occasionally have the word *prove* sprinkled throughout otherwise robust studies—sometimes even in an article's abstract.[11]

But proofs are the language of mathematics, not science. Math is based on a closed system that can be contained and controlled, and once a theorem is proven in math, it cannot be unproven (unless an error was made in calculations).

Science is different. Science exists to be challenged as it is based on the human observation of nature as it exists in the world, documented in what we call data, which produces what we call evidence. Over time, evidence will support some theories (with more evidence) over others (with less evidence). But nothing is ever proven in science; any theory can be overturned if enough evidence to the contrary is collected.

Furthermore, most if not all scientific data reflect knowledge that is situated in a particular place, time, and context rather than objective truth. That doesn't mean science doesn't come to accurate understandings of the world. (Far from it. I doubt that too many people would want to test theories of gravity by jumping off the nearest building.) But although the laws of physics remain constant regardless of where they are tested, or who is testing them, the same cannot be said for many other branches of natural and social science, which can provide vastly different results based on choices made throughout the research process.

For example, last year I taught a research experiences course for undergraduates and took students to various marshes in New York City in different states of restoration to measure the above- and below-ground biomass. I was using a methodology common to urban ecology called quadrat surveys, which involve placing a set number of quadrats (square frames) in random spots in the marsh and having teams of students count and identify each plant in the quadrat. In theory, once the total number of plants is counted, we could then extrapolate the findings from the quadrats to get a sense of the biomass across the entire marsh.

The process sounds straightforward and like it would lead us to one objective truth about the state of biomass in the marsh, but with each step, there were countless variables to consider and human choices to be made. How many transects should be put down? At what interval should the quadrats be placed? How large should the quadrats be (and thus, how many individual plants would each contain)? What season of the year should we conduct the surveys? Should the surveys be done after a natural disaster, such as a hurricane? How do we deal with unforeseen barriers to the placing of a transect in a particular site (for example, once while doing a transect survey in an urban forest, one of our quadrats landed on a man who was sleeping). What do we do when the transect is dominated by one grass, with thousands of individual stems (measure them all, or measure just a given quantity)?

Even when using the most robust established methods in the field, results of the analysis, while reflecting the population sampled, are limited in how well they reflect other, nonsampled, populations. Thus, because the development of scientific consensus requires the accumulation of evidence from multiple studies, replication of experiments across different contexts, settings, and populations is very important.

In other words, any one dataset will always have biases and limitations. Herein lies the central challenge with the type of meta-analysis used in the Cochrane report on mask wearing mentioned above. Such studies require the researchers involved to make a series of choices and to do the best they can with the results. For example, when doing systematic reviews, the first choice is to decide which studies to include in the analysis. In the case of the Cochrane report, the authors excluded observational and lab-based studies; they only included randomized control trials (RCTs) because typically RCTs are considered the most robust approach to evaluating the effectiveness of a given intervention. Unfortunately, given the newness of COVID-19, there were only two published RCTs of mask wearing early in the pandemic, one of which had too many methodological limitations to say anything conclusive.[12] In addition, neither study looked directly at the effectiveness of actual mask use in reducing illness; rather, they only measured whether there was a difference of increased mask usage and reduced symptomatic SARS-CoV-2 infections among participants who were encouraged (but not required) to wear masks relative to those in control groups who were not given that encouragement.[13] In such a situation, it was a curious choice not to include observational studies, which had been included in prior iterations of the review (the report was the sixth review in the series).[14] As science historian Naomi Oreskes points out in a brief overview of the Cochrane report, "Many studies are not as rigorous as we would like, because the messiness of the real world prevents it. But that does not mean they tell us nothing."[15]

Interestingly, the more methodologically robust RCT on mask

wearing during the pandemic included in the Cochrane analysis did find a significant difference among the treatment and the control group in terms of COVID-19–related symptoms. But as it was just one study, the authors then made another choice—again, one that made logical sense—to include prepandemic studies (such as those that measured the transmission of influenza) in their analysis. There were major limitations that reflected the choices and challenges researchers had faced in these studies as well. For example, one of the papers included in the report, which accounted for a large percentage of the calculations used in the meta-analysis, reported on the transmission rates of respiratory viruses among Mecca pilgrims who slept together in large tents by allocating individuals to no-mask control and mask treatment groups. But the researchers of that study reported major problems in implementing the protocol: the majority of those in the treatment group only wore the masks some of the time, while those in the control also reported wearing masks some of the time. Unsurprisingly, researchers found no difference in transmission rates between the two groups and concluded that the "trial was unable to provide conclusive evidence on facemask efficacy against viral respiratory infections most likely due to poor adherence to protocol."[16]

Given the limited number of reliable RCTs on mask-wearing during the pandemic, perhaps the most curious choice was the one made by Jefferson (the lead author) to declare with confidence that the study found that masks likely "don't work."[17] Certainly, more research is needed, and there is a great deal of uncertainty as to the specific contexts in which masks can effectively reduce the transmission of airborne viruses. But the conclusions (or lack thereof) of the Cochrane report said more about the difficulty of conducting meta-analyses on topics where there is very little reliable data than anything useful about the effectiveness of wearing masks during a pandemic.

How We See Data

Being overly conclusive with our findings is not just inaccurate; it is a massive disservice to science. For one thing, doing so obscures the process through which scientific knowledge grows and matures and why some theories hold up to the test of time while others are replaced by new theories. Perhaps most importantly, it strips science of its greatest strength: that scientists must constantly update their conclusions in the face of new evidence.

Is it reasonable to expect lay members of the public to dig into the data and question the results of a given study, especially when that study comes from an institution as reputable as the Cochrane Institute? Some scholars have argued that what is needed is a more data-literate society, one that can see through misleading headlines. But that begs a further question: is data literacy the answer?

Early in the COVID-19 pandemic, data analysis and visualization were said by some to be having a breakthrough moment in terms of increasing public awareness and understanding of science. Major newspapers such as the *New York Times* developed COVID-19 sections of their websites that were loaded with interactive maps, tables, and charts. Local, state, and national governments developed COVID-19 dashboards, reporting raw data on numbers of cases, tests administered, hospitalizations, and deaths, and websites that aggregated data from multiple sources such as Our World in Data saw unprecedented rates of online traffic.[18] Given the novelty of the pandemic—and that many of us were cooped up in our homes with little to do—looking at spreadsheets and graphs suddenly had a strange kind of appeal.

Initially, this interest in the numbers was seen as a boon for increasing data literacy, which has long been seen as a major goal in science education. Data literacy, which refers to the ability to ability to understand, analyze, and communicate data, is not an ability that most people have. Although anyone can ask a research question and most people would be able to quickly learn data collection skills, working with and interpreting data often require a higher level of expertise and some familiarity with statistics. It is also cognitively challenging and thus typically demands a plentiful amount of both curiosity and patience.

But as the months progressed, some scholars began noticing that increasing public access to data was not necessarily leading to greater data literacy, and, in some cases, it was contributing to further misinformation. For example, in one widely circulated graph that charted COVID-19 cases alongside countries in Europe that required mask mandates, the countries *without* mandates (namely, Norway, Sweden, and Denmark) were shown to have fewer COVID-19 cases than those with mandates. The graph thus provides a clear visual to suggest that masks are not effective at reducing the spread of the coronavirus. But missing from the chart was any context explaining potential confounding variables, such as information about lockdowns, population density, or even different colors to differentiate between countries. This information was essential to the context, as the countries with no mask mandates and fewer cases also happened to be those with much lower population density than countries with mandates and rising cases. In other words, the graph didn't provide enough information from which to make a reasonable conclusion and thus could be easily misinterpreted.[19]

It wasn't just mask skeptics who were creating their own graphs from publicly available data. Weeks before the Centers for Disease Control and Prevention recommended the public use of masks, a pro-mask group called Masks Save Lives posted a graph that displayed trends of COVID-19 cases from various countries. The graph showed countries with lower mask use (for example, the United States) having far greater cases than countries with higher rates of mask use (for example, Singapore). But when the chart was re-created to account for differences in population (scaled at cases per 100,000 individuals), the differences between countries were far less pronounced.[20]

This war of graphs boggled the minds of some data experts, who were quick to argue that it pointed to a bigger problem with a lack of data literacy.[21] On the face of things, that thought seemed to make sense. A more data-literate public would be able to understand, for example, the importance of providing context and disaggregating data to avoid the types of misinterpretations described in the previous examples. Such an educated constituency would easily and automatically recognize double or collapsed vertical axes, so went the argument.

A deeper exploration of this phenomena, however, found that the problem was less about mistakes and more about people making different choices regarding what data to include or exclude, something trained scientists do all the time. One study from a multidisciplinary team based at the Massachusetts Institute of Technology (MIT) found that many of the graphs and figures produced by antimaskers were virtually indistinguishable in appearance from those made by mainstream science sources.[22] Often, the numbers were the same; what was different was the analysis. Further qualitative research found that far from eschewing data, these groups were discussing—often with great nuance—how the data was collected and the types of biases that were inherent in the numbers. They were also encouraging one another to do their own analyses and make their own graphs and figures. They organized online tutorials to teach basic data interpretation skills, such as how to access, download, and work with large datasets in Excel and other programs. The authors of the MIT study thus concluded: "These findings suggest that the ability for the scientific community and public health departments to better convey the urgency of the US coronavirus pandemic may not be strengthened by introducing more downloadable datasets, by producing 'better visualizations' (e.g., graphics that are more intuitive or efficient), or by educating people on how to better interpret them.

This study shows that there is a fundamental epistemological conflict between maskers and anti-maskers, who use the same data but come to such different conclusions."[23]

In other words, different communities interpret data through their own preexisting worldviews. One of my favorite studies illustrating this finding was led by Dan Kahan, whose research focuses on how our cultural identities interfere with our reasoning skills. The study in question involved more than a thousand participants, all of whom were asked to solve a math problem that involved the same data (from a fictitious dataset). The participants were split into two groups. In the control group, participants were told that the data corresponded to the effectiveness of a new skin cream. In the treatment group, participants were told that the data related to whether a specific gun control law led to an increase or decrease in crime. Kahan and his colleagues found that in the skin cream control group, the only reliable predictor of whether a participant correctly solved the math problem was whether they were good at math.[24]

For the treatment group, who believed that the data referred to the effectiveness of gun control laws, the results were entirely different, however. Rather than one's numerical skills being the main predictor of whether they answered correctly, the determinant was personal ideology. If a participant's ideological views were in line with what the data said, they got the math problem right. If, on the other hand, the data went against their views, they tended to get the problem wrong. Even more interestingly, rather than reducing misinterpretations of data, the researchers found that higher levels of numeracy actually increased the chance that a given individual would let their political bias get in the way of mathematical reasoning. This study, and others like it, have provided convincing evidence that increasing access to and understanding of data aren't the silver bullets needed to address misinformation and distrust in science.[25]

Certainty, Not Uncertainty, Fuels Science Denialism

If data literacy isn't the answer, what is? How do we communicate uncertainty in science, and why is it important to do so?

One perhaps counterintuitive answer suggests that rather than attempting to persuade someone that the evidence is irrefutable, a more effective approach might be to find common ground on uncertainty. Talking about uncertainty can build credibility, as it admits that science doesn't definitively "prove" anything or have all the answers and encourages people to ask questions and have conversations. Furthermore, it forces people to admit what they also don't know, which is a key tool in challenging science denialism.[26]

To understand better, we must first unpack what we mean when we talk about uncertainty. Importantly, uncertainty is not the same as ignorance; rather, it is an awareness of one's own ignorance. When we are uncertain, we are aware of the limits of our knowledge.[27] Uncertainty is what fuels scientific research; science can always be disproven, and scientists are obliged to update their thinking in light of new evidence. We are all human and thus inclined to cling to outdated beliefs, but scientists who do so will eventually be on the fringes of scientific consensus and, if they persist in their beliefs, will find themselves straying into the world of pseudoscience (as some discredited climate science skeptics have done). A researcher may get something wrong, and that is forgivable, but if that same researcher insists they are right, they limit their ability to learn. It is this mentality of openness, humility, and growth that we can share with the public—that science isn't about a body of facts, but rather a community of practice that is always trying to explore the weaknesses in our own arguments.

Framed in this way, uncertainty can be a powerful tool for motivating more deliberate and thoughtful consideration of an issue. Being forced to reflect on the extent to which we don't understand something can make our brains more open to deliberation and reflection rather than to "gut-based" thinking.[28] This point is important because one consistent finding among science skeptics is that, compared to the larger public, they are typically more confident in their (inaccurate) beliefs. Research has found that not only are people who hold the most extreme antiscientific consensus views the least factually knowledgeable, but they perceive their levels of knowledge to be higher than average.[29] In other words, they're not just wrong—they're convinced they're right.[30] For example, one study that looked at understanding of and perceptions of genetically modified organisms found that individuals most opposed to genetically modified foods were not only the least knowledgeable about science and genetics, but they also rated their understanding of such issues as the highest.[31] Another study surveyed more than three thousand individuals regarding their objective and subjective knowledge levels on seven different issues for which there is scientific consensus (such as the big bang theory) and similarly found that participants with the highest levels of opposition to scientific consensus were the least knowledgeable and also the most overconfident in their knowledge.[32]

These studies are interesting not only for explaining psychological processes behind science skepticism, but also in terms of providing potential solutions for bursting unwarranted bubbles of overconfidence. Research in this area suggests that when people are made aware of their own relative lack of knowledge on a given topic, they can develop a more realistic assessment of what they do and don't know and may be more open to learning.[33] Simply by asking people to explain how specific mechanisms work can start this process. For example, many people have opinions on the mRNA technology used in the development of the COVID-19 vaccine, but few would be able to explain how exactly mRNA vaccines encode a transmembrane SARS-CoV-2 spike protein and then bind such proteins to angiotensin-converting enzymes (incidentally, I have no idea what this means—I just paraphrased it from an article I found online). Unless someone truly is an expert on a given topic, a request to explain something in a clear way will typically cause them to recognize how little they personally understand something. For most people (although notably not all), realizing they don't have all the answers will make them feel a bit humbled and can open the door to increased curiosity, which matters because higher levels of scientific curiosity have been found to be associated with less polarized views.[34] One study found that those who were more "scientifically curious" were also more open to learning and reading articles that challenged their preexisting views, whereas the less curious typically opted to read articles that agreed with what they already believed.[35]

Furthermore, some research has found that when uncertainty is communicated effectively, it can communicate a sense of transparency and honesty and can contribute to a greater understanding of both the method and findings. David Spiegelhalter, a statistician at the University of Cambridge and author of multiple books on probability and statistics, insists that people are far more able to understand uncertainty and risk than most scientists recognize. He further argues that aim of doing so should not be to get the public to trust science, but rather to demonstrate that scientists are trustworthy, explaining that

> it means being balanced, giving both sides of the argument. It means not trying to manipulate someone's emotions to make them do something or think in a certain way. It means actually informing them and helping them understand an issue, raising their understanding, and allowing them, empowering them to make a better choice. It means being upfront about uncertainty and whether the evidence is very good or not. . . . And just because you don't know everything doesn't mean you don't know anything. You need to say what you do know confidently about what you do know, and then you need to be confidently

humble. You say, well, but we don't know this. We're doing our best.[36]

There's a flip side to uncertainty, however. Although increasing one's sense of certainty in their own knowledge can make them less likely to fall into extreme positions, it can also be used to forestall science-based action. For example, in the 1960s, the tobacco companies infamously used doubt and uncertainty to discredit the scientific consensus about the link between smoking and lung cancer, a tactic that was later adopted in other antiscience campaigns, such as the fossil fuel industries' war on climate science. In his book *Doubt Is Their Product*, David Michaels wrote: "Industry has learned that debating the *science* is much easier and more effective than debating the *policy*."[37] "In field after field, year after year, conclusions that might support regulation are always disputed. Animal data are deemed not relevant, human data not representative, and exposure data not reliable."[38]

Because of the potential for doubt to increase science skepticism, science educators and communicators are often reluctant to acknowledge the limits of scientific knowledge, which perhaps is a main reason the word *prove* is often thrown around in science education. Science educators may be concerned that admitting uncertainty will lead students to be confused as to whether they can trust scientific conclusions. Scientists themselves are often similarly conflicted when speaking with the media or policy makers, and there is an active debate among the science communication community about the extent to which addressing uncertainty helps or hinders one's message.[39] Thus effectively communicating scientific uncertainty in a world in which bad actors may use it against us is a significant challenge and an area where more research and training on effective techniques is sorely needed.[40] Some research has suggested it may be useful to distinguish between uncertainty in scientific consensus (for example, scientists don't agree) and the uncertainty in a particular study (for example, explaining the significance of error bars). Although the former has been linked to increasing public skepticism on science-related issues, the latter has been found to increase trust and credibility in science more generally.[41]

Teaching Uncertainty

Another approach to addressing this challenge is to teach uncertainty as a key feature—not a bug—of scientific thinking. In 1996, the visionary and prolific Carl Sagan wrote: "It is a supreme challenge for the popularizer of science to make clear the actual, tortuous history of its great discoveries and the misapprehensions and occasional stubborn refusal by its practitioners to change course. Many, perhaps most, science textbooks for budding scientists tread lightly here. It is enormously easier to present in an appealing way the wisdom distilled from centuries of patient and collective interrogation of Nature than to detail the messy distillation apparatus. The method of science, as stodgy and grumpy as it may seem, is far more important than the findings of science."[42] As Sagan aptly suggested, it is hard to teach this "stodgy and grumpy" method of science through static and passive knowledge-sharing vehicles, such as textbooks. But some scientists have been trying to change the narrative by offering courses that put uncertainty at the heart of what drives science.

A course created by surgery professor Marlys H. Witte of the University of Arizona in the 1980s called "Introduction to Medical and Other Ignorance" was one of the first such offerings. Initially, Witte struggled to find a home for the course, but she argued that it was needed to help students understand that the value of medical science wasn't in teaching what was already known, it was teaching at what still needed to be discovered. A similar course was created many years later at Columbia University by neuroscientist Stuart J. Firestein, author of the book *Ignorance: How It Drives Science*. He created his course after realizing his students were graduating with the mistaken impression that everything important was already known in the field of neuroscience, which he partly attributed to the 1,414-page textbook they were assigned to read for one of his courses. He wrote: "This crucial element in science was being left out for the students. The undone part of science that gets us into the lab early and keeps us there late, the thing that 'turns your crank,' the very driving force of science, the exhilaration of the unknown, all this is missing from our classrooms. In short, we are failing to teach the ignorance, the most critical part of the whole operation."[43]

This point is often missing when we neglect to bring others into the process through which we make discoveries. In his TED Talk, physicist Uri Alon used the metaphor of a cloud to describe the fuzziness of discovery, where hypothesis A does not lead to result B but rather to unhelpful and exasperating outcome after outcome, undermining our faith in our ability to do science. He called this fuzziness "the cloud" and said that only when we have spent sufficient struggling with the often-confusing processes of scientific discovery do we stumble upon C—an unexpected and exciting finding worth writing about. We do not tell others of our stumbling in the dark; rather, we write up the analysis as if we went from A to C in a linear and straightforward fashion.[44]

But doing so belies the importance of our time spent in the cloud, Alon said, because the cloud represents the frontier of science—the space where the known blends into the unknown, and that's where the most important discoveries are often made. He argued that teaching students about the difficulties—and importance—of this stage will lead to fewer people quitting science as they begin to see the struggle as a key part of the journey. Acknowledging what happens in the cloud also requires us to admit to the emotional and subjective aspects of the journey that we all engage in when we do science. He said: "You see, science seeks knowledge that's objective and rational. That's the beautiful thing about science. But we also have a cultural myth that the doing of science, what we do every day to get that knowledge, is also only objective and rational, like Mr. Spock. And when you label something as objective and rational, automatically, the other side, the subjective and emotional, become labeled as non-science or anti-science or threatening to science, and we just don't talk about it."[45]

Talking about it can also lead to more interest in and more positive attitudes toward science and scientists. For example, one study exposed students to stories about science through different treatments: one that shared stories about three different scientists' struggles, one that shared stories about these same scientists' lifetime achievements, and one that only shared information about the physics contents the students were studying. The research found that the stories about achievements did not result in any increases in students' interest in the content, recall of concepts, or improvement in scores, but rather led to increased negative perceptions of scientists and science. In contrast, the stories about scientists struggling increased students' interest in science, increased their understanding of science concepts, improved scientific problem solving, and, furthermore, led to their having more positive views of scientists. Thus the researchers concluded that it wasn't necessarily having just increased awareness of scientists that improved perceptions and, in fact, that more information about their achievements strengthened preexisting stereotypes about scientists and science. Rather, it was "when students are given opportunities to learn more about scientists' struggles, they are more likely to see scientists as ordinary people who encounter challenges and struggle in their scientific discoveries."[46]

The Importance of Sharing Our Values

A key theme in the examples above is the importance of transparency and honesty. Transparency is crucial for establishing and strengthening (and sometimes rebuilding) public trust in science. In a survey of US adults, respondents were given a list of items and asked to what extent each mattered to them when deciding to accept a scientific finding. The items were whether the study has been published in a peer-reviewed science journal, whether the scientists make their data and methods available and are completely transparent about their methods, and whether the scientists involved in the study disclose the individuals and organizations that funded their work.[47] Interestingly, the survey found that the most important factor was whether the scientists make their data and methods available and were transparent about methods. Of second importance was also a matter of honesty and transparency: whether the scientists involved in the study disclosed the individuals and organizations who funded their work.

Thus it is important that researchers be honest, don't exaggerate who they are or what they can do, and are up front about their values when communicating science to the public (and their intentions for doing so). When the consequences of our research have implications for society—whether in terms of public health, environment, economic policies, or social debates—it is inevitable that we, as researchers, will have some biases in the matter.[48] It does no one any good to pretend that we are unbiased when we have skin in the game—we will only succeed in fooling ourselves. As Naomi Oreskes argued, "Value neutrality is a tinfoil shield."[49] Oreskes suggested that rather than trying to downplay our values, we need to bring the values of other communities into conversation with science. To do so, we must recognize that some scientific debates are more about the values that direct the science than the science itself, and that even the decision about how much uncertainty to communicate, and how to communicate such uncertainty, is value-based.[50]

For example, going back to the face mask debates mentioned earlier in this chapter, ultimately the decision about whether to wear a face mask (or to enforce mask mandates) comes down to one of values. Do masks have the potential to limit the spread of disease? The research, although inconclusive and with some caveats (how they are worn, the type of fabric used), suggests they do.[51] Do mask mandates have potential negative consequences for society? The research, again inconclusive and context dependent, finds that can be the case (for example, in terms of learning for students and communication).[52] Ultimately, the decision is not just one based on science, but also of balancing trade-offs based on values.[53] Thus to say that wearing a mask during a pandemic is a scientifically based behavior is accurate *and also* incomplete because it neglects the very real trade-offs made in terms of public health, social cooperation, and learning. In the end, such decisions are not about "following the science," but instead must be about what is best for a particular social context.[54] As one *New York Times* reporter wrote, "There is no one correct answer to our Covid dilemmas. People are going to disagree passionately, and that's frequently how it should be. Most policy options have both benefits and drawbacks."[55]

If I could go back in time to that classroom in March 2020, I would tell my students I didn't have any good answers about what was happening with the pandemic. And although that might not be what they wanted to hear, it would have shown them that it is okay to admit when we are uncertain.

CHAPTER 10

In the Belly of the Beast: Scientists, Policy-Making, and Advocacy

IN 2017, MORE THAN ONE MILLION SCIENTISTS and their allies convened in Washington, DC, and satellite locations around the world for first-ever March for Science. Although framed by organizers as nonpartisan, the march was partly a reaction to the antiscience rhetoric and actions taken by the new administration in the United States, which during Donald Trump's first few weeks of becoming US president had scrubbed all references to climate change, appointed a climate denier as the administrator of the Environmental Protection Agency, and pledged significant cuts in government funding for scientific research.[1] March-goers held signs that included "Science Improves Decisions," "Scientists Pursuing Truth, Saving the World," and "Science Serving the Common Good" and participated in chants such as "When science is under attack, what do we do? Stand up, fight back!"

In many ways, the March for Science was unprecedented in the history of science. Since the Enlightenment, scientists have taken pains to steer clear of advocacy due to concerns that such engagement could damage one's credibility or "objectivity" as a scientist.[2] For example, at an American Association for the Advancement of Science meeting in 1993, a reporter for *Science* said that he had put several prominent scientists, including Carl Sagan, E. O. Wilson, and Bruce Ames, on a "blacklist" for their increasing role in advocacy.[3] Others have argued that such engagement goes beyond the scope of science because of the premise that science and politics must be kept separate.[4] In 2020, *Nature* published a three-part podcast series called *"Stick to the Science": When Science Gets Political*, to explore public backlash to a story it had published about the damage Trump was doing to science. "Keep opinions out of a science page. This page should be about studies with empirical data!" read one social media post. "Politics should not feature in *Nature*'s aims and scope," said another.[5] And in 2023, an earth scientist was fired from her research post at the Oak Ridge National Laboratory in Tennessee for engaging in political actions to raise awareness about climate change.[6]

However, even as scientists are told to stay out of politics, they are increasingly encouraged to try to influence policy-making.[7] In other words, there is often a distinction made between directly advocating for a particular political course of action and providing evidence related to the issue at hand. Surveys have found that a large majority of scientists believe that they should play an active role in public debates around specific science issues, such as stem cell research, and that most scientists see policy makers as the most important group with whom to engage.[8] But the distinction between politics and policy-making is open to multiple interpretations. The first March for Science in 2017 and its subsequent iterations are perhaps the most visible sign that the line is increasingly being blurred. For example, conservative news sources, such as Fox News, derided the 2018 march as an "attack on Trump and Republicans," pointing out that march-goers carried anti-Trump signs and that headline speakers featured prominent Democratic politicians.[9] Surveys later showed that march-goers were overwhelmingly politically active and that 72 percent identified as Democrats compared to only

2.5 percent who identified as Republicans.[10] Some march participants even later ran for office; Jasmine Clark, who has a PhD in microbiology from Emory University and directed the March for Science Atlanta, was elected to the Georgia House of Representatives in 2018.

The blurring of lines between political advocacy and science-based contributions to policy-making have been of increasing concern to many scientific organizations, who fear public backlash and a broader loss of public trust if doing science is seen as a political act. These concerns, while not new, raise to the surface long-debated questions about the roles that scientists can (and perhaps should) play when they attempt to influence policy-making with their research.[11] Scholars who have written about these roles suggest that engaging in policy-making is important work that should be done carefully and thoughtfully to ensure that scientists' integrity is not weakened. However, engaging in policy-making can mean many things, and it can be challenging (and frustrating) for researchers to navigate the (often contradictory) flood of advice about how to best do so, let alone take the initial steps of undertaking the work itself.[12] Does one stick to the facts, or is it better to voice one's opinion? Should scientists try to meet with policy makers directly, or should they work through intermediaries?

These debates illustrate that attempting to influence policy with science requires both awareness and intentionality, similar to any other aspect of scientific research. In other words, it means making conscious choices, and what determines such choices will be not only a matter of personal interest and conviction, but also one of professional status—one's level of seniority, for example, and the extent to which one feels they are able to take risks in their career. These choices, like others in research, will be both deeply personal and also context dependent. And perhaps most crucially, they require a clear-eyed understanding of what one is getting into when they seek to influence policy with their research. To aid with this understanding, this chapter provides an overview of what is meant by the policy-making process, how some researchers have chosen to engage in it, and some of the opportunities and costs of doing so. Although this chapter focuses on examples from the United States, some of the lessons may be useful for researchers working in other contexts.

Policy-Making Is Not Just One Thing

Michelle Land is a professor of environmental policy and law at Pace University and a registered lobbyist in New York State. Since 2014, Land and her colleague John Cronin—also a Pace professor—have been training students to research and lobby local and regional jurisdictions to address environmental problems through an undergraduate course called the Environmental Policy Clinic.[13] The clinic's agenda has included banning microbeads in personal care products, training road salt applicators to protect sensitive ecosystems, and protecting the welfare of animals used for entertainment. The students spend months researching a particular issue, understanding perspectives of diverse interest groups, and strategizing about policy design and then travel to meet with policy makers and make their case.

In December 2023, Land and her students celebrated a major victory when Kathy Hochul, New York State governor, announced her signature on the bill to ban wildlife killing contests, one of the issues students had spent years working on. Wildlife killing contests are organized events in which participants compete (often for cash prizes) to kill as many of the target animal as possible within a set period—usually a few days. The contests are often defended as a (misguided) attempt to control the population of a species (often midsized to large predators, such as foxes or bobcats) that are considered so-called pests. Because there is no limit on the number of animals that can be killed, such contests can temporarily decimate local wildlife populations. Aside from moral and ethical concerns associated with the contests, research has found that indiscriminate hunting is not an effective way to control populations of popularly hunted species, such as coyotes.[14] Although the Humane Society of the United States spearheaded the drafting of the bill, Land and her students played an integral role in getting the bill passed into law through years of research, advocacy, and campaigning. Their experiences shed light on challenges and opportunities for researchers engaging in advocacy to influence policy decisions.

One of the first things Land explained to me about the policy process is that it's multifaceted, with different avenues through which decisions are made. Engagement takes place at numerous levels across diverse venues and involves different types of actors. Thus researchers must be cognizant of who, where, and why they might engage, even before attending to the how. Sometimes policy is made at central levels of government, such as by federal, state, or regional branches. But other times (and perhaps increasingly), central government passes the buck for enacting policy to other entities, including agencies, and even private industry.[15] Different venues have different rules and cultures, which makes the processes through which policies are made context dependent.

Thus understanding policy requires getting a handle on a larger landscape within which there are multiple actors alternately working together, in tandem, or in conflict. Such a landscape can be extremely complex and dynamic, especially at the more visible levels of politics. For example, in the United States, one of the seemingly most "straightforward" ways to enact policy is to convince legislators to sponsor and shepherd the passage of a specific bill. Legislators may introduce dozens of bills during a given session but only have the capacity to push one through the system. Furthermore, even if the bill is prioritized, the sponsor must convince their fellow chamber members to vote on its passage. To be helpful to a bill sponsor, one must understand legislative process, identify key actors, solve problems raised with bill language, and negotiate competing priorities. It is also important to know when to move on. If a policy issue is not getting traction in one arena, it may be better to shift venues rather than waste time and resources engaging with someone not likely to be persuaded by your proposed solution. As Land explained to me, "Part of the strategic planning is asking: at what level of governance do you approach this initially? Where do you start? And then where do you pivot to if that's not productive?"[16]

Because of this high level of complexity and flux, lobbyists often work as part of a broader coalition of organizations and interest groups, each contributing a key piece of the puzzle. Just as policy is not made in a vacuum by one person (in democratic societies, anyway), neither is attempting to amplify the role of evidence in policy-making a one-person operation. In other words, to be effective, such advocates understand that they cannot go it alone. Teaming up can be challenging, however. Although members of a coalition may have shared interests, they might not agree on how to tackle a given policy issue. Some advocates might aim to raise as much public awareness as possible about a given issue to garner support, whereas others may prefer an approach that operates mainly outside the public arena to prevent the mobilization of opposition groups. Being part of a coalition means recognizing and negotiating between different perspectives and tactics and understanding one's personal role in the bigger picture.[17]

For example, in the case of the wildlife killing contest bill, Land and her students wanted to raise public awareness about the issue and garner support early in the legislative cycle, but others in the coalition preferred to avoid creating noise out of concern that drawing attention would pull opponents out of the weeds and lead to backlash. Land and her students adapted. They limited their advocacy to within the Pace University community, doing research and gathering more than five hundred signatures from students, faculty, and staff for a petition in support of the law. Once the bill was moving through relevant legislative committees, they were ready. They traveled to the state capital to deliver the petition and meet with lawmakers. As Land described: "The students in the program created a campaign strategy and learned how to implement that strategy effectively. . . . They researched the issue, created social media platforms that addressed the issue, and reached out to other students for signatures. They needed to convince some people, but the good lesson was they learned how to engage people on controversial topics."[18]

The Woman behind the Curtain

If science is a process largely misunderstood by policy makers, so too is policy-making a process not well understood by scientists. Thus it stands to reason that the individuals—the who—at the center of this process are least understood of all. For those of us outside this process, we might be inclined to see policy makers as endowed with superhuman powers. They are the ones, after all, who make the decisions that affect the rest of us. They decide whether to implement a carbon tax that would curb use of fossil fuels, they issue vaccine mandates, and they pass or cut budgets on science and technology spending.

Due to the important role that policy makers play in the process, well-intentioned researchers often dedicate time to advocating and trying to influence them directly. They may call, write to, or even visit elected officials. But when attempts to communicate evidence do not lead to desired change, scientists sometimes place the blame on policy makers for not heeding their advice.[19] This reaction is understandable, but it's also potentially misplaced. The evidence of evidence-based policy-making has largely found that approaches that aim to "get the right evidence into the right hands" are not very effective on their own even when they do an objectively good job at packaging the facts in brief, accessible, and attractive formats.[20] In an article that synthesizes the evidence of what works in communicating with policy makers, policy scholar Paul Cairney and organizational psychologist Richard Kwiatkowski sought to upend the "fairytale that heroic scientists are thwarted by villainous politicians drawing on their emotions and deeply held beliefs in a 'post truth' world." They argued that "repeatedly stating the need for 'rational' and 'evidence-based policy-making' is pointless, and naively 'speaking truth to power' counterproductive."[21] Rather, they argued that what is needed is how to understand the social, cultural, and political landscape of policy-making, and the policy makers themselves are important elements in the process.

For example, a common misunderstanding that researchers who are new to the policy-making process may have is that policy makers are not interested in, or aware of, scientific evidence. I've heard that myself many times. But policy makers are not tasked with completing a synthesis of the evidence-based literature and making logical decisions based only on what the science says. They must also listen to constituents, be aware of fiscal limitations, and consult broader public opinion. Thus, although policy makers often seek to take scientific findings into account, they must be able to balance multiple information sources, including differing perspectives on given information.[22] They cannot make decisions without considering how those decisions will be perceived by their constituents or the elected officials who appointed them to their posts (who are in turn evaluated by their constituents). So, many times policy makers are forced to take public positions that don't necessarily reflect their personal views. It can be especially evident with politically controversial issues such as climate change, where a given policy maker might acknowledge the scientific consensus in private but is not willing to advance the issue through public policy. Furthermore, policy makers frequently make decisions in high-pressure, high-stakes environments, which are associated with more emotion and value-based (as compared to purely "logical") decision-making. Such environments often require "quick and dirty" decision-making rather than a careful and deliberate consideration of the evidence.[23] In other words, policy makers are flawed humans, just like the rest of us, situated in contexts that are not ideal for optimal decision-making.

One event that upends the notion that policy makers are not interested in science is Evidence Week, which is organized by Sense about Science (mentioned in chapter 7) to discuss the role of evidence in policy in the United Kingdom. Evidence Week is extremely popular with policy makers, attracting more than a hundred members of Parliament, Lords, and parliamentary staff each year.[24] Policy makers and researchers interact through trainings, three-minute policy briefings, and panel discussions, all of which have been shown to reveal new domains in which research can influence policy. For example, in one session in 2022, policy makers engaged with software that demonstrated the impact of aerodynamics on fuel consumption and spoke with key scientists on implications for legislation.[25] Policy interventions around fuel efficiency tend to focus on gasoline economy, which determine subsidies for hybrid or electric vehicles. But the opportunity to engage with researchers working on the design stage of vehicles pointed to a new policy intervention point—how to create policies to incentivize more aerodynamic design.[26]

Another common misunderstanding among researchers new to the policy-making process is that policy makers are a homogeneous group. Dr. Rebecca Adler Miserendino is an environmental scientist with two decades of experience working at the intersection of science and policy in the United States. When I asked her whether "calling a policy maker" was good advice, she said the following:

> I don't think that advice is bad, but it's incomplete. Not all policy makers are equal in terms of their ability to receive and act on information or recommendations provided to them by stakeholders. For example, if you are interested to change US law, it will be important to know what committees your senator or representative serves on, what issues your

member championed in the past, and what your member's current priorities are. All this information will help you target your message to resonate with the member and their staff and increase the likelihood that your outreach can make an impact. If you seek to engage with the executive branch, you will need to know which agency officials oversee the policy matters that connect to the work you are doing. Ideally, you should also have a sense of at which level your outreach will be most effective. Engaging with the policy process is not as straightforward as "just reach out to your delegate."[27]

Similarly, the most visible faces of policy-making are not necessarily those with whom you need to speak. Policy makers include many types of civil servants, as well as their advisors, who can be very influential in how decisions get made. Land told me that being willing to "be helpful" was some of the best advice she could give to someone interested in advocacy and policy and emphasized the importance of developing relationships with staff in policy makers' offices. She explained that it was essential to develop a relationship to understand communication style—when a bill was worth trying to push through as opposed to when it would be wasted effort. For example, she had developed a certain relationship with one staffer and knew that, when there was interest, he was quick to respond to emails. But when the topic was a no-go, he would be silent, which was a message that let her know it was not a good time for the specific issue to be addressed, and even if she didn't understand why, to respect his silence. She explained:

> Most people believe meeting with the legislator is the ultimate goal. The truth is, meeting with a staff member, legislative director or committee counsel can often be more informative and productive because their job is to know everything about everything. The legislator at the end of the day is going to lean on the staff and the counsel about details of a bill's impact and viability. I mean, yes, it's driven by values of

the legislator themselves, but they have a lot of other things to think about. . . . So, having that relationship whereby the staff is willing to be forthright about things and give you the insider angle reflects your quality of engagement. They're not obligated to do that, but if they trust you and they value you and you've been helpful to them, they're going to guide you when you need it.[28]

Recently, I was at a science-policy symposium organized by graduate students at Pennsylvania State University. The last session of the day featured two speakers involved in evidence-based policy-making at the state level, one of whom was Jordan Harris, a member of the Pennsylvania House of Representatives. The speakers were asked how they engage with experts when making decisions, and Harris gave the example of a bill that had recently come up about social workers. He said that as part of the process of trying to understand the issue, he reached out to two social workers whom he knew personally and whose contact information he had stored on his phone. This story, and the examples provided by Miserendino and Land, demonstrate the importance of building of relationships that are long-term, with two-way dialogue and trust. Land put it as follows:

> I think the biggest one is that people think it's like a one-off thing. Researchers think, "I have this issue. I want to make my opinions heard." And so they have the meetings (with policy makers) and then they go home exasperated feeling like "they did nothing about it." But it's absolutely not a one-off thing. People think if they are just heard and somehow effective enough in that meeting, the outcome will be successful. It's a long-term engagement, multiple years, multiple angles that you have to take. You have to be helpful.[29]

One approach to assess whether you've done the groundwork of trying to influence policy is whether you would pass the 7-11 test from

Research!America.[30] The test is simple: if you go to a convenience store and run into a policy maker, do you know them? Do they know you? If so, you're in the game. If not, perhaps you have more preparatory work to do.

Surfing the Swamp

Using evidence to influence policy is sometimes described as being a surfer, waiting patiently for the next big wave.[31] Both the political environment and the timing of sharing evidence need to line up. Windows of opportunity can open and close quickly, which is another reason advocates work as part of a collective that have their ears to the ground on various fronts. In this sense, it's often less a matter of using persuasive evidence and more one of knowing who to prod and when. Being attuned to windows of opportunities is a particular challenge for evidence-based policy, as policy makers and researchers work on vastly different time scales. Although much research operates on long time frames, with some projects taking years for the accumulation of sufficiently robust data, policy makers often must act in fast-paced, high-stakes environments that change from season to season.[32] However, some policy issues—particularly controversial ones—can take decades to gain traction. The bipolar pace of policy-making can be alternatingly stressful and frustrating for researchers, who may see windows of opportunity close if they do not feel ready to provide evidence if it comes with a high degree of uncertainty. As pointed out in chapter 9, though, uncertainty can be communicated in productive ways, and it may be more important to share results—even preliminary ones—at the right time if such evidence has the potential to be relevant to a policy being deliberated.

One of the most commonly cited examples of science-based policy-making is the Montreal Protocol, which was the result of a decade-long effort to reduce ozone-depleting chemicals. Part of the story often forgotten is that the scientific evidence at the time was largely incomplete. For example, scientists didn't yet understand regional variations in the depletion of the ozone layer and disagreed on the reductions needed.[33] But the policy window, largely created by public concern, was open, and those who drafted the Montreal Protocol did so with the awareness that both the science and the policy would evolve together. Rather than providing a set of objective facts from which policy emerged, political scientist Karen Litfin explained, "the political impact of scientific knowledge [was] determined far more by its incorporation into larger discursive practices than by either its validity or the degree to which it [was] accepted by scientists."[34]

Although policy windows are often viewed as serendipitous, they also have the potential to be harnessed through strategic approaches.[35] In other words, researchers don't have to sit around and wait for a policy window to open, but instead can take more active steps. Framing is one approach that can be used to put a given issue onto a policy makers' radar by telling narratives that resonate with prior assumptions and specific values.[36] For example, in many countries, framing health-related policy solutions in terms of economic savings can be more powerful than focusing solely on the benefits to patients, especially as competing interest groups will typically use such cost-based arguments.[37] However, one challenge with framing is the need to understand the motivations and concerns of the target audience, which can shift depending on other priorities. Furthermore, framing can backfire if the person receiving the message doubts the authenticity of the messenger and suspects that they are trying to manipulate them.[38]

Land told me that one approach she has found to be effective is to share one's personal perspective. For example, when her students met with policy makers about the wildlife killing contest bill, they discussed their feelings of anger for what they saw as the mismanagement of wildlife, which they saw as a betrayal of the government's role. One student was a hunter, from a family of hunters, and Land said that his perspective was particularly valuable. He told policy makers that he believed the contests conveyed the wrong message about hunting—that fair chase doesn't matter—and that wildlife is disposable. Land said, "If you just come in and recite the facts, they could read an article about that. But if you can connect the facts to stories and people and their constituents—you've got to tie the thread through all of those pieces—and then it becomes compelling and effective."[39]

An additional consideration is that policies do not typically emerge from nothing; rather, often they are fixes to problems with previous policies. Miserendino told me that researchers who seek to influence policy-making tend to be idealistic, thinking about what a better way it would be to do x or y. She said: "It is relatively rare for policy makers to create a new policy from scratch. Instead, policy makers must work within the boundaries that are provided to them by current law and practice—they cannot completely reinvent the system (even if they made grandiose claims about 'draining the swamp'). As such, they need solutions that take into consideration what exists and to work from there to make modifications or additions."[40]

All Roads Lead to Rome

Sometimes making evidence-based policy may mean not engaging directly with policy makers at all. Policy makers are strongly influenced by their constituents, especially those who represent groups. If an issue is facing a lot of conflict, policy makers will wait to see how differing groups battle it out—and to understand which groups have more power, especially during election seasons—before making related policy decisions. It can mean working to inform and educate members of interest groups to get them on board, which can reduce concerns that the policy issue will lead to a politically divisive, and potentially career-ending, fight.

For example, one of the most controversial policy issues in the Pacific Northwest of the United States has been how to manage the region's old-growth forests. In the 1980s, intense conflict between loggers and environmentalists led to both violence and lawsuits on both sides. In the early 2000s, a handful of loggers and environmentalists fed up with the constant battles created the Blue Mountain Forest Partners, a collaborative of diverse actors focused on healthy forests and forest communities in the Blue Mountains of Oregon. Since its founding, the collaborative has played a large role in policy-making, creating and putting forward "zones of agreement" reports on proposed pathways for forest management and policy.[41] The collaborative uses scientific data to inform decision-making, which provides many opportunities for the engagement of researchers in how policy is ultimately made. One area of concern has been how to best design salvage logging to minimize negative impacts for threatened species of woodpeckers. In collaboration with the collaborative, wildlife ecologist Victoria Saab conducted years of studies of the impacts of salvage logging on woodpecker abundance. Her research is being used to inform how the timber companies carry out salvage logging, addressing the habitat needs of species like woodpeckers when considering what trees to cut through proposed "wildlife habitat zones of agreement."[42]

Being involved in this type of approach requires the ability to tolerate and accept imperfect solutions rather than seeking ideal ones. For example, Saab's research found that one species of woodpeckers, black-backed woodpeckers, was found not to be able to tolerate any level of logging, which pointed to the importance of having some areas of forest where no logging at all was permitted.[43] The proposed zones of agreement are also not ideal for the timber companies, which would make more money by salvaging all the timber on the ground. But it is often said that "politics is the art of compromise," which is especially true of agreements developed as part of a coalition of diverse interest groups. The work of the Blue Mountain Forest Partners also demonstrates the importance of navigating conflict and difference when incorporating conflicting perspectives and values. The collaborative was established not because the loggers and environmentalists wanted to work together, but because they realized they had no choice.[44]

The power of indirect approaches to influencing policy is visible in research by behavioral scientists Anuj Shah, Oeindrila Dube, and Sandy Jo MacArthur at the University of Chicago Crime Lab. As part of their research, they worked with the Chicago Police Department to develop a training program to reduce adverse outcomes in policing, which include things like excessive force and unnecessary arrests. In an interview, Shah said that being aware of the timing and the local context was essential for the project to be successful; the researchers had approached the Chicago Police Department at a time when there was both local and national concern about policing and a departmental recognition of the need for more targeted training.[45] It was also essential to ensure that the training was designed in a way that worked for the officers via a process that was specific to the rules and norms of the department. As Shah described, "We worked iteratively with them to create a training program that was responsive to the context,"[46] and their efforts paid off: the researchers found that officers who received the training were less likely to engage in discretionary arrests, particularly of Black civilians, and were also less likely to be injured on duty.[47] However, the research also found that the beneficial effects of the training wore off after several months, pointing to the importance of long-term investments in evidence-based training efforts.

The above examples show some of the benefits of researchers engaging directly with people who make or influence policy. But researchers interested in this type of work need to understand one important caveat: although science can inform policy, it will not be the sole determining factor. Policy-making is ultimately about how to balance different types of information, values, and contexts. In other words, it is about trade-offs. Sometimes a decision will be made that seemingly doesn't make sense if one looks only at the scientific evidence. But perhaps that decision was made as part of a compromise for getting something else done. Researchers interested in walking this path should try not to get too discouraged when things don't go their way but to persevere by continuing to building trust and accountability until a better opportunity presents itself.[48] Also, by engaging with policy makers over a longer period of time, a researcher may be able to have an indirect impact on policy decisions by influencing the ways certain issues are framed and debated.[49] Being aware that seeking to influence policy is a long game, with multiple players, can provide a healthy set of expectations for researchers in terms of deciding if and how they wish to be involved.

To Engage or Not to Engage

One of the most surprising pieces of advice I heard from people with experience in policy-making is if researchers attempt to influence policy without being aware of how it works or who the key players are, they can sometimes do more harm than good. Pushing hard on something when there isn't a receptive audience for it yet creates a reactive type of culture of engaging in the science-policy process rather than an intentional and deliberate one. It is important to be aware that advocating in the wrong way at the wrong time can derail the policy process, and those present must be aware of what an effective approach would look like.

Thus individuals working at the science-policy interface must work within defined bounds to ensure that they are credible to those they seek to influence.[50] Because of this need for legitimacy, some have suggested that researchers seeking to influence policy should determine in advance what type of role they wish to play to avoid unintended consequences that may arise if they advocate for certain positions without first disclosing their values.[51] There is a danger of hiding value-based policy arguments behind scientific evidence, and particularly with controversial issues.[52] Also, engaging in advocacy can have high costs for researchers, which are not necessarily distributed equally. In particular, researchers who come from communities historically underrepresented in science, whether due to race, ethnicity, gender, or "ableness," are at higher risk of negative consequences to their careers or even to their personal safety.[53] Furthermore, political situations in some places around the world can pose additional risks for scientists who wish to influence policy with their research. For example, researchers in Brazil whose scientific work directly or indirectly challenged official government policy under the Bolsonaro administration (2019–2022) frequently came under attack, forcing some to flee the country due to death threats and other forms of intimidation.[54]

Because of the deep level of expertise needed to engage effectively, trying to influence policy may feel more like a career choice than an event. A friend of mine shared that in his field of physics there is a clear distinction between theoretical physicists, who work on abstract problems related to the nature of the universe, and those paid to solve specific industry problems, such as how to more make photovoltaic cells more efficient. Although the industry physicists frequently draw on theoretical science, they are more involved in putting out patents and less involved in writing peer-reviewed papers. This model, common also in engineering, points to the opportunity to create additional application-focused research positions in other fields, such as environmental science, sociology, and health.[55]

Indeed, what is called for are organizations, structures, and individuals who can effectively navigate a process rather than simply bridge divides. This idea is not new—"boundary organizations" and "knowledge brokers" have been around for decades.[56] Such organizations and individuals are set up to create consistent communication and interaction among science, policy, and practice through the development of long-term relationships and mutually valuable projects. What makes boundary organizations and knowledge brokers different are their accountability to actors in both science and policy and their goal of protecting science from becoming politicized. Moreover, they help make science information useful and on point. Boundary entities maintain the flow of information between different parties and focus on sustaining relationships, and because they navigate between research and political worlds on a daily basis, they must have a deep understanding of policy cycles as well as the scientific process.

One boundary organization, the Lenfest Ocean Program, seeks out key policy issues related to oceans and fisheries and provides medium-sized research grants designed to provide relevant scientific information on said issues.[57] The program also provides ongoing support throughout the research process, helping researchers draft information about salient points, create messaging to get the word out, follow the policy cycle to know when and to whom to present information and decide when to get the press involved. This approach also challenges the "fund and forget" model common to many research-granting agencies.[58]

Similarly, knowledge brokers are individuals who understand the values, cultures, and language of both scientists and policy makers and who move freely between different communities. For example, knowledge brokers working for the United Kingdom's Department for Environment Food and Rural Affairs found that involving individuals responsible for setting farming policy early in the research process led to interest in and uptake of research findings into decision-making. The researchers created opportunities through focus groups for policy makers to hear directly from farmers about their perspectives on given policy directions. This stepping outside the office enabled spaces of encounter where differing ideas and experiences could be exchanged, leading to the development of new policy ideas.[59]

Researchers interested in learning more about the policy process are in luck, as there are an increasing number of opportunities to learn about and engage more directly with the agencies and individuals who make policy. For example, the American Association for the Advancement of

Science offers one-year Science and Technology Policy Fellowships for researchers at multiple levels, including midcareer, to work in the federal government for a year.[60] In Europe, various science-policy think tanks offer paid positions (both short and long term) for researchers interested in contributing to European Union policy.[61] And across the world, nonprofit organizations and advocacy groups serve as boundary organizations that can provide guidance and support for researchers seeking to influence policy with their science. In some places, scientific societies engage in advocacy related to their topic of interest. For example, the American Institute of Biological Sciences' Public Policy Office provides training for scientists interested in communicating with policy makers.[62] Thus a good first step for many researchers interested in influencing policy is to spend some time identifying relevant opportunities in their particular location and area of interest. One rule of thumb for getting started is to assume that there is a group of people who care about the same thing you do and have already done some groundwork on the issue.

In 2024, Land celebrated her tenth year running the policy clinic, which has engaged more than seventy students across more than twenty-five disciplines of study. Some of her students now have careers in policy and advocacy and work with agencies and organizations across the Unites States.[63] This success points to the role of higher education to teach the next generation of researchers about the policy-making process as an integral part of their training,[64] and it can also help provide alternative career paths for scientists given that there are far more job candidates with PhDs than tenure-track openings.[65] As Land said, "The clinic's success relies on its reputation as a consistent, trustworthy purveyor of good information to decision makers, requiring deep understanding of legislative and regulatory processes. Perhaps the most important lesson learned by our student clinicians is that policy advocacy is an iterative, multistakeholder, long-term pursuit for which there is no quick fix substitute."[66]

CONCLUSION AND ACKNOWLEDGMENTS

From Boldly Going to Steadily Engaging

Since discovering the *Star Trek* universe a few years ago, I've eagerly explored its various series: *Deep Space Nine*, *Voyager*, and my new favorite, *Lower Decks*. This last series is something of an unlikely addition to the *Star Trek* canon—it's a cartoon parody, and its main characters are the underlings of Star Fleet society on a relatively unimportant starship (the USS *Cerritos*). The crew of the *Cerritos* specialize in "second contact," the less-glamorous task of following up with alien societies after the intrepid first encounter was established by a different set of explorers (such as Captain Jean-Luc Picard). As described in the first episode of the series, "We get all the paperwork signed, we make sure we're spelling the name of the planet right, and we get to know all the good places to eat."[1] In other words, second contact is necessary, but not very prestigious, work. It means checking up on the societies where more famous explorers first "went boldly" and then promptly forgot about as they moved on to more exciting discoveries.

In this way and others, *Lower Decks* provides a subtle critique of *Star Trek*'s mission to "discover new life-forms and civilizations." This critique is meaningful and has its equivalent in science, which tends to emphasize the deeds of the heroic pioneers of new theories, discoveries, and inventions. Historian Lauren Beck refers to this tendency as "firsting," described as "the process through which a scholar presents an act, circumstance, or phenomenon generated by man, or accomplishment to have occurred for the first time."[2] As Beck and others pointed out, firsting both glorifies certain individuals and knowledge systems by diminishing, or even erasing, the contributions of others.[3] Firsting is part and parcel of our current academic system because researchers are required to demonstrate our unique—and *individual*—contributions to science to advance in our careers.[4]

But as detailed throughout this book, extra academic impact in science is not a one-and-done individual affair. It requires not just contact with people, places, and ideas, but attention to the *nature of the contact*.[5] It is dependent on an understanding of communication as a process of two-way dialogue based in listening and connection, and long-term commitments to people, places, and ideas. It is an undertaking of the many rather than the mission of a select few. And because impact travels in multiple directions, there are no small parts; it is an all-hands-on-deck effort. That's what I like most about *Lower Decks*. By focusing on the exploits of the lower-ranking crew members of a relatively unimportant starship, *Lower Decks* demonstrates that the real day-to-day work of keeping the Federation both peaceful and prosperous takes much more than the bold, new discoveries of *Star Trek*'s prestigious explorers. Seeing missions through the eyes of those tasked with cleaning the carbon filters reminds us of the many ways to contribute to a world in which science truly does benefit all.

Those of us in the scientific community might be wise to take note. When we think of "what counts" in demonstrating the impact of our research, it is clear what is expected: scholarly publications in top journals, large research grants, and getting our science in the news (the bigger the media outlet, the better). We are also keenly aware of what doesn't often count: asking practitioners, community groups, and policy makers what research questions they would find important, presenting our research at a local church, and getting to know staffers in policy makers' offices. But as I hope this book has made clear, it is not just the big shiny discoveries that make science matter, but also the less glamourous activities that require long-term engagement and follow-through that can lead to major societal impacts.

I interviewed dozens of people for this book, and at the end of each interview I asked the interviewee if they had a recommendation for something I needed to include. Almost without fail, people talked about the lack of incentives and financial support to do the work that they knew mattered. For example, Dr. Roma Solomon, retired director of the Core Group Polio Project, told me that the most challenging part of the work was not getting skeptical parents to get their children vaccinated, but rather convincing those in positions of power about the importance of investing in outreach, a challenge that remains today. "What is sad now is that all the lessons that polio taught us have sort of been forgotten. The world seems to be forgetting because civil society organizations always come last," she said. "Why should they come last? Why should communication about the vaccine not be as important as manufacturing the vaccine?"[6]

So I will end this book with two calls. The first is directed to those in positions of authority and relative power in the scientific community: funders, scientific leaders, and academic administrators. Look at the evidence of what works to spread scientifically based ideas and behaviors and provide tangible support for those activities. Incentivize and fund activities that go beyond the "Please Just Listen to Me" model of science communication in favor of opportunities for researchers to connect with and build long-term relationships of trust with communities, practitioners, policy makers, and other interested groups. First contact is exciting, but second contact—and everything that comes after—is what will make for a different kind of future.

The second call is to my peers. Here I will refer again to my favorite *Star Trek* cartoon, which above all suggests that although the work of exploring the universe is serious stuff, it doesn't mean we must take ourselves quite so seriously. Academics are frequently characterized as lofty, boring, and exclusive because—truth be told—we often are. We write (and sometimes speak) in a language that is not only unintelligible to the average person, but that often makes communication with researchers from other disciplines challenging. We privilege some voices, ideas, and perspectives at the expense of others. Similarly, we pay too much attention to the activities of the great few at the top rather than the work of countless others who make science happen. But as science and technology scholars have long argued, there is no line in the sand between science and society.[7] Our brilliant ideas do not occur to us only from within ivory walls, but rather from a powerful constellation of sources. Thus one way we can expand who is part of the scientific community is by recognizing and giving credit to the neighbors, friends, nonfaculty coworkers, and students, among others, who regularly impact and give meaning to our work and lives.

Toward that aim, I have a long list of people and communities that I would like to thank for their contributions to this book. First, special thanks to my colleague, friend, and business partner, Monica Palta, who helped with this book in too many ways to count. I am also deeply grateful for those who took the time to speak with me about their own perspectives on the issues in this book, including Craig Allen, Matt Bomgardner, Nabanita Borah, Leslie Bowman, Emily Brown, Tracey Brown, Julia Byrd, Bill Chain, Molly Cheatum, María Eugenia Copa, Tim Crowhill Sauder, Maggie Douglas, Sara Perl Egendorf, Chelsea Foster, Denny Garman, Michelle Land, Crystal Lee, Max Liboiron, Burnell Martin, Violet Martin, Dave McLaughlin, Rebecca Adler Miserendino,

Liz Neely, Christina Pagel, Jiwan Palta, Donal Pérez Gutiérrez, Harrison Rhodes, Roger Rohrer, Anuj Shah, Roma Solomon, David Spiegelhalter, Earl Swift, Piyush Tantia, Vivian Tseng, and Tracey Van Kempen.

In a bit of a panic that this book would not go through traditional peer review, I put calls out on social media and listservs to find readers. And found them I did—more than a dozen experts across various fields and multiple countries were kind enough to take on the labor of reviewing sections of this book. Some of these readers ended up having stories of their own to share, and a few have become friends. Thanks go to Aimee Bernard, Grant Currin, Sharon Frank-Hirsch, Stephanie Flynn James, Megan Jones, Bindu Panikker, Rod Parnell, Ada Roseti, Divya Sharma, Gary Silverman, and Teresa Spezio. A very special acknowledgement to Jane Wolfson, who read and gave detailed and critically constructive comments on all twelve chapters of the book. Thanks also go to the online communities where I found these generous souls, including the Association of Environmental Studies and Science listserv and the *Science of Science Communication* and the *Academic Writing—Reciprocal Feedback* Facebook communities. Thanks also go to the *Star Trek Shitposting* Facebook community for suggesting some *Lower Decks* episodes.

I am especially grateful to other colleagues who reviewed parts of this book, especially John Besley, Kevin Elliott, Scott Peacor, Mark Reed, and Jennifer Shirk. My readers also included family members and friends, who let me know in frank terms what resonated and what didn't. Thanks go to Christine Gregg, Scott Markle, Mark Thielking, Abby Toomey, and Michael Toomey.

I am also appreciative of people and communities who informed this book indirectly, before I first sat down to write. I am grateful to Matthew Aiello-Lammens, Rob Currie, Meg Domroese, Armando Medinaceli, Hernan Nay, Celín Quenevo, Eleanor Sterling, Mark Zsurka, the Takana, Tsimane', Mosetén, and Takana-Quechua Indigenous communities in Bolivia, and the park guards of Madidi National Park and the Pilón Lajas Biosphere Reserve. Several podcasts also informed ideas in this book, particularly Alan Alda's *Clear and Vivid*, David McRaney's *You Are Not So Smart*, and Shankar Vedantam's *Hidden Brain*.

I am grateful to the Pace University community, which is full of wonderful and caring colleagues, students, and staff. I wrote this book during my sabbatical year, which was an incredible privilege for me but also put more work on others. I especially thank Melanie Dupuis, Michael Finewood, Michael Rubbo, and Denise Santiago for picking up the slack and checking on me once in a while. My students have often driven me to think about the impact of being a professor in new ways, especially students like Camryn Becker and Tatyana Graham. I am also grateful to the wonderful librarians on both the New York City and the Pleasantville campuses who were so helpful throughout this process in obtaining countless interlibrary loans and often renewing them after the due dates had long since passed.

My deep thanks go to others who made this book a reality, including the entire Island Press team, especially to Erin Johnson, without whose unwavering support this book would not have happened. I also extend my sincere thanks to the Island Press production and marketing team, especially Sharis Simonian. I am also grateful to Kathleen Lafferty, who copyedited the book, and Emily Iskin, who created the art of the flowcharts on the book website, accessible at islandpress.org/impactful.

Finally, I would like to thank my parents, John Toomey and Eileen Toomey, whose encouragement (and perhaps a bit of overpraise) has helped me believe that I could and would. I especially thank my mother, who carefully fact-checked every single reference not once, but twice, who patiently formatted my notes section, and who sent me regular text messages, laden with the types of inspiring emojis that every writer (and daughter) needs. I love you so much.

Notes

Introduction. Science—The Next Generation

1. In this book, I use the terms *science* and *research* somewhat interchangeably to describe many different forms of gathering knowledge about the natural and social world that are based on the rigorous collection of data and assessment of evidence. For a thorough read on the debate of what constitutes science, see C. A. Taylor, *Defining Science: A Rhetoric of Demarcation* (University of Wisconsin Press, 1996).
2. For example, in the United States, see "Major Gaps between the Public, Scientists on Key Issues," Pew Research Center, July 1, 2015 (https://www.pewresearch.org/internet/interactives/public-scientists-opinion-gap/). Referring to perspectives on vaccines from around the world, see H. J. Larson, C. Jarrett, E. Eckersberger, D. M. D. Smith, P. Paterson, "Understanding Vaccine Hesitancy around Vaccines and Vaccination from a Global Perspective: A Systematic Review of Published Literature, 2007–2012," *Vaccine* 32, no. 19 (2014): 2150–59. Referring to perspectives on the shape of Earth in the United States, see L. Hamilton, *Conspiracy vs. Science: A Survey of U.S. Public Beliefs* (University of New Hampshire Carsey School of Public Policy, 2022) (https://carsey.unh.edu/publication/conspiracy-vs-science-a-survey-of-us-public-beliefs). And in Brazil, see C. Orsi, *Survey: Beliefs in Science and Pseudoscience in Brazil* (Instituto Questão de Ciência, 2019) (https://www.revistaquestaodeciencia.com.br/english/2019/05/23/survey-beliefs-science-and-pseudoscience-brazil).

3. Creationism in schools in the United States and Canada: J. R. Wiles, "Overwhelming Scientific Confidence in Evolution and Its Centrality in Science Education—and the Public Disconnect," *Science Education Review* 9, no. 1 (2010): 18–27. Decline in vaccination coverage: R. L. Eagan, H. J. Larson, and A. de Figueiredo, "Recent Trends in Vaccine Coverage and Confidence: A Cause for Concern," *Human Vaccines and Immunotherapeutics* 19, no. 2 (2023): 2237374.
4. As quoted in *Merchants of Doubt*, directed by R. Kenner, Sony Pictures Classics, 2014. Quote by Hansen begins at time stamp 21:08.
5. For a good read on this idea, see M. Saadia, *Trekonomics: The Economics of Star Trek* (Pipertext, 2016).
6. For a more nuanced discussion of how disabilities are represented in *Star Trek*, see G. K. Abercrombie-Winstanley and A. M. Callus, "Disability in Intergalactic Environments: The Representation of Disability Issues in *Star Trek*," *New York Review of Science Fiction* 28, no. 8 (2016): 4–9.
7. Too often, researchers in the "hard" sciences, such as physics, biology, and chemistry, have butted heads with social science and humanities scholars. That is particularly the case in discussions of ethics, power issues, and objectivity in science, which are topics of focus in science and technology studies, anthropology, philosophy, and the history of science, among other disciplines. For example, the "Science Wars" of the 1980s and 1990s were a series of contentious discussions between scholars in both "camps" about the role of science in society.
8. For another interesting exploration of what constitutes impact in research, see J. Bayley, *Creating Meaningful Impact: The Essential Guide to Developing an Impact-Literate Mindset* (Emerald Publishing Limited, 2023).
9. Here I mean that I've taken an active stand against the use of jargon throughout the book. Too often have I submitted papers to journals and been required to include and cite the fashionable terms of the day in revisions rather than just writing what I mean in clear language that anyone can read. See V. Clayton, "The Needless Complexity of Academic Writing: A New Movement Strives for Simplicity," *The Atlantic*, October 26, 2015. https://www.theatlantic.com/education/archive/2015/10/complex-academic-writing/412255
10. The book website is accessible at islandpress.org/impactful.
11. M. Gasman, and T. Nguyen, *Making Black Scientists: A Call to Action* (Harvard University Press, 2019); L. M. P. Munoz, *Women in Science Now: Stories and Strategies for Achieving Equity* (Columbia University Press, 2023).
12. K. C. Elliott, *A Tapestry of Values: An Introduction to Values in Science* (Oxford University Press, 2017).
13. Another limitation of this book, related to my positionality, is the overreliance on studies of human attitudes and behaviors that were conducted in WEIRD

settings (western, educated, industrialized, rich, and democratic). See J. Henrich, S. J. Heine, and A. Norenzayan, "Most People Are Not Weird," *Nature* 466, no. 7302 (2010): 29.
14. Approximately 35 people from six different countries provided feedback on various sections of the book; most people reviewed only one or two chapters, but each chapter was reviewed by at least five people.
15. *Star Trek: The Next Generation*, "When the Bough Breaks," season 1, episode 17, directed by K. Manners, aired February 15, 1988. Quote begins at time stamp 30:15.

Chapter 1. Will You Please Just Listen to Me?

1. *Don't Look Up*, directed by A. McKay, released December 5, 2021, on Netflix.
2. A. Katwala, "*Don't Look Up* Nails the Frustration of Being a Scientist," *Wired*, December 20, 2021. https://www.wired.com/story/dont-look-up-climate-scientists/
3. As quoted in J. Burton, "How 'Don't Look Up' Got the Science Scarily Accurate," *Newsweek*, December 20, 2021. https://www.newsweek.com/dont-look-science-accurate-dr-amy-mainzer-1662550
4. This conceptualization is similar to that of the "deficit model" in science communication, which is based on the premise that if only people understood science (had higher levels of scientific literacy, for example), they would not only support science and technology but would also be supportive of policy recommendations based in scientific findings. Decades of research across multiple fields (such as behavioral economics, policy studies, and science and technology studies) have found that the deficit model does not accurately depict how science communication works. However, many scientists continue to see it as central to science communication efforts. See M. J. Simis, H. Madden, M. A. Cacciatore, and S. K. Yeo, "The Lure of Rationality: Why Does the Deficit Model Persist in Science Communication?," *Public Understanding of Science* 25, no. 4 (2016): 400–414.
5. *Don't Look Up*. Scene begins at time stamp 1:05:28.
6. "Tangier Island Historic District," Virginia Department of Historic Resources, accessed March 20, 2014. https://www.dhr.virginia.gov/historic-registers/309-0001/
7. Earl Swift, personal communication, December 15, 2023. See also "Tangier Island Historic District."
8. D. M. Schulte, K. M. Dridge, and M. H. Hudgins, "Climate Change and the Evolution and Fate of the Tangier Islands of Chesapeake Bay, USA," *Scientific Reports* 5, no. 1 (2015): 1–7.

9. "Tangier, Va.," DATA USA, 2021. https://datausa.io/profile/geo/tangier-va#demographics
10. Z. Wu and D. Schulte, "Predictions of the Climate Change-Driven Exodus of the Town of Tangier, the Last Offshore Island Fishing Community in Virginia's Chesapeake Bay," *Frontiers in Climate* 3 (2021): 779774.
11. J. Portnoy, "Why Sen. Kaine Wants to Save Trump Country from Sinking into the Chesapeake," *Washington Post*, September 2, 2017. https://www.washingtonpost.com/local/virginia-politics/why-sen-kaine-wants-to-save-trump-country-from-sinking-into-the-chesapeake/2017/09/02/906d0016-8695-11e7-961d-2f373b3977ee_story.html
12. Vice News, "This Virginia Island Is Literally Sinking into the Sea," aired November 7, 2016. https://www.youtube.com/watch?v=FKih6nRw2d8
13. CNN, "Rising Seas May Wash Away This US Town," aired June 9, 2017. https://www.youtube.com/watch?v=lBPiTsibWQg. The story was sympathetic to the islanders' plight but also mentioned that more than 87 percent of Tangier's vote in the 2016 presidential election went to Trump.
14. *The Late Show with Stephen Colbert*, "Trump Says 'Not to Worry' about Rising Sea Levels," June 17, 2017. https://www.youtube.com/watch?v=o5AxKZF_xP8
15. *Full Frontal with Samantha Bee*, "What's Happening to Tangier Island?," November 15, 2017. https://www.youtube.com/watch?v=WZoVYl9ltcA
16. N. Raihani, *The Social Instinct: How Cooperation Shaped the World* (Random House, 2021).
17. An insightful book that describes this in more detail is S. A. Sloman and P. Fernbach, *The Knowledge Illusion: Why We Never Think Alone* (Penguin, 2018).
18. S. Gorman and J. Gorman, *Denying to the Grave: Why We Ignore the Facts That Will Save Us* (Oxford University Press, 2016).
19. The term *cognitive miser* was popularized by Susan Fiske and Shelley Taylor in 1984. See S. T. Fiske and S. E. Taylor, *Social Cognition: From Brains to Culture* (Sage, 2013).
20. A classic book on heuristics is D. Kahneman, *Thinking, Fast and Slow* (Macmillan, 2011).
21. N. Oreskes, *Why Trust Science?* (Princeton University Press, 2019). For example, the dominance of string theory in physics is a classic example of where individual and collective biases within a scientific community are said to have halted decades of progress in the field. See L. Smolin, *The Trouble with Physics: The Rise of String Theory, the Fall of a Science, and What Comes Next* (Houghton Mifflin Harcourt, 2007).

22. See Jay Van Bavel's extensive work on social identity and group dynamics. For example, J. Van Bavel and D. Packer, *The Power of Us: Harnessing Our Shared Identities to Improve Performance, Increase Cooperation, and Promote Social Harmony* (Little, Brown, 2021).
23. P. Wallisch, "Illumination Assumptions Account for Individual Differences in the Perceptual Interpretation of a Profoundly Ambiguous Stimulus in the Color Domain: 'The Dress,'" *Journal of Vision* 17, no. 4, art. 5 (2017): 1–14.
24. D. McRaney, *How Minds Change: The Surprising Science of Belief, Opinion, and Persuasion*, (Oneworld Publications, 2022).
25. D. M. Kahan, D. A. Hoffman, D. Braman, D. Evans, and J. J. Rachlinski, "They Saw a Protest: Cognitive Illiberalism and the Speech Conduct Distinction," *Stanford Law Review* 64 (2012): 851–906.
26. McRaney, *How Minds Change*, 83.
27. Gorman and Gorman, *Denying to the Grave*.
28. K. M. Bellizzi, "Cognitive Biases and Brain Biology Help Explain Why Facts Don't Change Minds," *The Conversation* (August 11, 2022). https://theconversation.com/cognitive-biases-and-brain-biology-help-explain-why-facts-dont-change-minds-186530
29. E. Swift, *Chesapeake Requiem: A Year with the Watermen of Vanishing Tangier Island* (HarperCollins, 2018).
30. Swift, *Chesapeake Requiem*, 219.
31. C. Vaughn, "Tangier Mayor Disputes Cause of Island's Land Loss on CNN's Al Gore Town Hall," *Daily Times*, August 2, 2017. https://www.delmarvanow.com/story/news/2017/08/02/tangier-mayor-questions-al-gore-islands-land-loss-cnn-town-hall/532023001/
32. N. Oreskes and E. M. Conway, *Merchants of Doubt: How a Handful of Scientists Obscured the Truth on Issues from Tobacco Smoke to Global Warming* (Bloomsbury Publishing USA, 2011). The film (of the same name) is also a great resource.
33. C. Battig, "Heartland Chicago Climate Meeting," June 6, 2010. https://climateis.com/2010/06/06/heartland-chicago-climate-meeting/
34. R. E. Dunlap and A. M. McCright, "A Widening Gap: Republican and Democratic Views on Climate Change," *Environment: Science and Policy for Sustainable Development* 50, no. 5 (2008): 26–35.
35. S. Byrne and P. S. Hart, "The Boomerang Effect a Synthesis of Findings and a Preliminary Theoretical Framework," *Annals of the International Communication Association* 33, no. 1 (2009): 3–37.
36. Which increased more than twenty points, from 37 percent to 59 percent. See Dunlap and McCright, "A Widening Gap."

37. Vaughn, "Tangier Mayor Disputes Cause of Island's Land Loss."
38. Earl Swift, Zoom interview, November 21, 2023.
39. Vaughn, "Tangier Mayor Disputes Cause of Island's Land Loss."
40. The average household carbon footprint in Tangier is 39.9, compared to 49 as the US average. See C. Jones and D. M. Kammen, "Spatial Distribution of US Household Carbon Footprints Reveals Suburbanization Undermines Greenhouse Gas Benefits of Urban Population Density," *Environmental Science and Technology* 48, no. 2 (2014): 895–902. Also see the map tool at https://coolclimate.berkeley.edu/maps
41. "Al Gore: Other Resources," *Nobel Prize*, accessed April 10, 2024. https://www.nobelprize.org/prizes/peace/2007/gore/other-resources/
42. N. Popovich, "Climate Change Rises as a Public Priority. But It's More Partisan Than Ever," *New York Times*, February 20, 2020. https://www.nytimes.com/interactive/2020/02/20/climate/climate-change-polls.html
43. Pew Research Center, *As Economic Concerns Recede, Environmental Protection Rises on the Public's Policy Agenda*, February 13, 2020. https://www.pewresearch.org/politics/2020/02/13/as-economic-concerns-recede-environmental-protection-rises-on-the-publics-policy-agenda/
44. "Trends in Atmospheric Carbon Dioxide (CO2)," National Oceanic and Atmospheric Administration, Global Monitoring Laboratory. https://gml.noaa.gov/ccgg/trends/global.html accessed April 15, 2024
45. M. Donoghoe, A. M. Perry, S. Gross, E. Ijjasz-Vasquez, J. B. Keller, J. W. McArthur, S. Patnaik, et al., "The Successes and Failures of COP28," December 14, 2023. https://www.brookings.edu/articles/the-successes-and-failures-of-cop28/
46. *Don't Look Up*, directed by A. McKay, released December 5, 2021, on Netflix. Dr. Mindy's speech starts around time stamp 1:30.

Chapter 2. Will I Please Just Listen to You?

1. "Estimated Polio Cases" (dataset), World Health Organization (original data), processed by Our World in Data, accessed April 10, 2024. https://ourworldindata.org/polio; In 2009, India had 741 cases out of 1,604 cases globally. "Wild Poliovirus Cases" (dataset), World Health Organization (original data), processed by Our World in Data, accessed April 10, 2024. https://ourworldindata.org/polio
2. A. Bellatin, A. Hyder, S. Rao, P. C. Zhang, and A. M. McGahan, "Overcoming Vaccine Deployment Challenges among the Hardest to Reach: Lessons from Polio Elimination in India," *BMJ Global Health* 6, no. 4 (2021): e005125.
3. R. Koul, "Polio Eradication: India Offers to Help Pakistan," *Biospectrum*,

October 27, 2014. https://www.biospectrumindia.com/news/95/4915/polio-eradication-india-offers-to-help-pakistan.html

4. "History of the Polio Vaccine," World Health Organization, accessed March 27, 2024. https://www.who.int/news-room/spotlight/history-of-vaccination/history-of-polio-vaccination
5. R. Solomon, "Involvement of Civil Society in India's Polio Eradication Program: Lessons Learned," *American Journal of Tropical Medicine and Hygiene* 101, no. 4 suppl. (2019): 15.
6. Currently, this role is undertaken formally by Accredited Social Health Activist workers. See "About Accredited Social Health Activist (ASHA)," Ministry of Health and Family Welfare, Government of India, accessed April 13, 2024. https://nhm.gov.in/index1.php?lang=1&level=1&sublinkid=150&lid=226
7. Dr. Roma Solomon, Zoom interview, January 4, 2024.
8. For example, one very hard to reach population included migrant families, so collaboration with these groups led to the identification of areas where these families would be working at different times. Lack of regular employment led entire villages to migrate to a certain region of the country for seasonal work in the brick kiln industry. Health workers held workshops with brick kiln owners to obtain lists of children eligible for the vaccine, which helped ensure these children were accounted for. See Bellatin et al. "Overcoming Vaccine Deployment Challenges."
9. H. Jafari, "Polio-Free India: It Seemed Impossible until It Was Done," World Health Organization, Polio Eradication Initiative, accessed April 13, 2024. https://www.emro.who.int/polio-eradication/news/polio-free-india-it-seemed-impossible-until-it-was-done.html
10. N. Deutsch, P. Singh, V. Singh, R. Curtis, and A. R. Siddique, "Legacy of Polio—Use of India's Social Mobilization Network for Strengthening of the Universal Immunization Program in India," *Journal of Infectious Diseases* 216, no. suppl_1 (2017): S260–66.
11. Solomon, interview.
12. R. M. Wolfe and L. K. Sharp, "Anti-Vaccinationists Past and Present," *BMJ* 325, no. 7361 (2002): 430–32.
13. A. S. V. Shah, C. Gribben, J. Bishop, P. Hanlon, D. Caldwell, R. Wood, M. Reid, et al., "Effect of Vaccination on Transmission of SARS-COV-2," *New England Journal of Medicine* 385, no. 18 (2021): 1718–20.
14. "Immunization Coverage," *World Health Organization*, July 18, 2023. https://www.who.int/news-room/fact-sheets/detail/immunization-coverage
15. P. Plans-Rubió, "Vaccination Coverage for Routine Vaccines and Herd Immunity Levels against Measles and Pertussis in the World in 2019," *Vaccines* 9, no.

3 (2021): 256; and P. Fine, K. Eames, and D. L. Heymann, "'Herd Immunity': A Rough Guide," *Clinical Infectious Diseases* 52, no. 7 (2011): 911–16.

16. T. Marima and S. Nolen, "More Than 700 Children Have Died in a Measles Outbreak in Zimbabwe," *New York Times*, September 24, 2022. https://www.nytimes.com/2022/09/24/health/measles-outbreak-zimbabwe.html

17. C. Gowda and A. F. Dempsey, "The Rise (and Fall?) of Parental Vaccine Hesitancy," *Human Vaccines and Immunotherapeutics* 9, no. 8 (2013): 1755–62.

18. D. A. Gust, N. Darling, A. Kennedy, and B. Schwartz, "Parents with Doubts about Vaccines: Which Vaccines and Reasons Why," *Pediatrics* 122, no. 4 (2008): 718–25.

19. H. J. Larson, C. Jarrett, E. Eckersberger, D. M. D. Smith, and P. Paterson, "Understanding Vaccine Hesitancy around Vaccines and Vaccination from a Global Perspective: A Systematic Review of Published Literature, 2007–2012," *Vaccine* 32, no. 19 (2014): 2150–59.

20. M. Grandahl, S. C. Paek, S. Grisurapong, P. Sherer, T. Tydén, P. Lundberg, "Parents' Knowledge, Beliefs, and Acceptance of the HPV Vaccination in Relation to their Socio-Demographics and Religious Beliefs: A Cross-Sectional Study In Thailand," *PLOS ONE* 13 (2018): 1–17.

21. R. Trepanowski and D. Drążkowski, "Cross-National Comparison of Religion as a Predictor of COVID-19 Vaccination Rates," *Journal of Religion and Health* 61 (2022): 2198–211; E. Lahav, S. Shahrabani, M. Rosenboim, and Y. Tsutsui, "Is Stronger Religious Faith Associated with a Greater Willingness to Take the COVID-19 Vaccine? Evidence from Israel and Japan," *European Journal of Health Economics* 23, no. 4 (2022): 687–703.

22. Religious beliefs play a strong role in HPV acceptance in many Asian countries: L. P. Wong, P.-F. Wong, M. M. Hashim, L. Han, Y. Lin, Z. Hu, Q. Zhao, and G. D. Zimet, "Multidimensional Social and Cultural Norms Influencing HPV Vaccine Hesitancy in Asia," *Human Vaccines and Immunotherapeutics* 16, no. 7 (2020): 1611–22. Studies of parents in Lebanon have found no association between HPV acceptance and religious beliefs; see R. Zakhour, H. Tamim, F. Faytrouni, M. Makki, R. Hojeij, and L. Charafeddine, "Determinants of Human Papillomavirus Vaccine Hesitancy among Lebanese Parents," *PLOS ONE* 18, no. 12 (2023): e0295644.

23. J. D. Grabenstein, "What the World's Religions Teach, Applied to Vaccines and Immune Globulins," *Vaccine* 31, no. 16 (2013): 2011–23.

24. M. Klymak and T. Vlandas, "Partisanship and COVID-19 Vaccination in the UK," *Scientific Reports* 12, no. 1 (2022): 19785.

25. L. Silver and A. Connaughton, "Partisanship Colors Views of COVID-19

Handling Across Advanced Economies," Pew Research Center, August 11, 2022. https://www.pewresearch.org/global/2022/08/11/partisanship-colors-views-of-COVID-19-handling-across-advanced-economies/

26. A. L. Wagner, N. B. Masters, G. J. Domek, J. L. Mathew, X. Sun, E. J. Asturias, J. Ren, et al., "Comparisons of Vaccine Hesitancy across Five Low- and Middle-Income Countries," *Vaccines* 7, no. 4 (2019): 155; N. Bergen, K. Kirkby, C. V. Fuertes, A. Schlotheuber, L. Menning, S. Mac Feely, K. O'Brien, and A. R. Hosseinpoor, "Global State of Education-Related Inequality in COVID-19 Vaccine Coverage, Structural Barriers, Vaccine Hesitancy, and Vaccine Refusal: Findings from the Global COVID-19 Trends and Impact Survey," *Lancet Global Health* 11, no. 2 (2023): e207–17.

27. Among parents in Saudi Arabia: S. S. Alsubaie, I. M. Gosadi, B. M. Alsaadi, N. B. Albacker, M. A. Bawazir, N. Bin-Daud, W. B. Almanie, M. M. Alsaadi, and F. A. Alzamil, "Vaccine Hesitancy among Saudi Parents and Its Determinants: Result from the WHO SAGE Working Group on Vaccine Hesitancy Survey Tool," *Saudi Medical Journal* 40, no. 12 (2019): 1242. Among parents in the United States: F. Wei, J. P. Mullooly, M. Goodman, M. C. McCarty, A. M. Hanson, B. Crane, and J. D. Nordin, "Identification and Characteristics of Vaccine Refusers," *BMC Pediatrics* 9 (2009): 1–9.

28. See page 1768 (and related references) in E. Dubé, C. Laberge, M. Guay, P. Bramadat, R. Roy, and J. A. Bettinger, "Vaccine Hesitancy: An Overview," *Human Vaccines and Immunotherapeutics* 9, no. 8 (2013): 1763–73.

29. For example, one study identified 147 factors that impact decisions about whether to accept, delay, or refuse a vaccine. See H. Larson, J. Leask, S. Aggett, N. Sevdalis, and A. Thomson, "A Multidisciplinary Research Agenda for Understanding Vaccine-Related Decisions," *Vaccines* 1 (2013): 293–304. See also Larson et al., "Understanding Vaccine Hesitancy."

30. M. J. Goldenberg, "Public Misunderstanding of Science? Reframing the Problem of Vaccine Hesitancy," *Perspectives on Science* 24, no. 5 (2016): 552–81.

31. M. J. Goldenberg, *Vaccine Hesitancy: Public Trust, Expertise, and the War on Science.* (University of Pittsburgh Press, 2021).

32. It is also important to distinguish between the behavior of trust (making oneself vulnerable) and trustworthiness perceptions (ability, benevolence, integrity). See J. C. Besley and L. A. Tiffany, "What Are You Assessing When You Measure "Trust" in Scientists with a Direct Measure?," *Public Understanding of Science* 32, no. 6 (2023): 709–26.

33. J. Brownlie and A. Howson, "'Leaps of Faith' and MMR: An Empirical Study of Trust," *Sociology* 39, no. 2 (2005): 221–39.

34. B. Misztal, *Trust in Modern Societies: The Search for the Bases of Social Order* (Wiley, 2013), 18.
35. K. Bardosh, A. de Figueiredo, R. Gur-Arie, E. Jamrozik, J. Doidge, T. Lemmens, S. Keshavjee, J. E. Graham, and S. Baral, "The Unintended Consequences of COVID-19 Vaccine Policy: Why Mandates, Passports and Restrictions May Cause More Harm Than Good," *BMJ Global Health* 7, no. 5 (2022): e008684.
36. R. D. Silverman and L. F. Wiley, "Shaming Vaccine Refusal," *Journal of Law, Medicine and Ethics* 45, no. 4 (2017): 569–81.
37. Some research finds that positive emotional messages, such as altruism and hope, can help encourage vaccination. See W. S. Chou, and A. Budenz, "Considering Emotion in COVID-19 Vaccine Communication: Addressing Vaccine Hesitancy and Fostering Vaccine Confidence," *Health Communication* 35, no. 14 (2020): 1718–22.
38. K. Estep and P. Greenberg, "Opting Out: Individualism and Vaccine Refusal in Pockets of Socioeconomic Homogeneity," *American Sociological Review* 85, no. 6 (2020): 957–91.
39. E. K. Brunson, "The Impact of Social Networks on Parents' Vaccination Decisions," *Pediatrics* 131, no. 5 (2013): e1397–404.
40. S. Ritzen, "Aziz Ansari Likens the Aaron Rodgers Debate to 'Making Fun of the Quarterback for Doing Bad on the Science Test,'" *We Got This Covered*, video, January 26, 2022. https://wegotthiscovered.com/news/aziz-ansari-ridicules-aaron-rodgers-nightclub-comedian-netflix/ (The special can be seen at https://www.netflix.com/title/81572737.)
41. K. Batelaan, "'It's Not the Science We Distrust; It's the Scientists': Reframing the Anti-Vaccination Movement within Black Communities," *Global Public Health* 17, no. 6 (2022): 1099–12.
42. Solomon, interview.
43. E. Dubé, D. Gagnon, and N. E. MacDonald, "Strategies Intended to Address Vaccine Hesitancy: Review of Published Reviews," *Vaccine* 33, no. 34 (2015): 4191–203.
44. B. Nyhan, J. Reifler, S. Richey, and G. L. Freed, "Effective Messages in Vaccine Promotion: A Randomized Trial," *Pediatrics* 133, no. 4 (2014): e835–42.
45. See similar findings from S. Pluviano, C. Watt, and S. Della Sala, "Misinformation Lingers in Memory: Failure of Three Pro-Vaccination Strategies," *PLOS ONE* 12, no. 7 (2017): e0181640.
46. T. Oraby, V. Thampi, and C. T. Bauch, "The Influence of Social Norms on the Dynamics of Vaccinating Behaviour for Paediatric Infectious Diseases," *Proceedings of the Royal Society B: Biological Sciences* 281, no. 1780 (2014):

20133172; J. Leask, H. W. Willaby, and J. Kaufman, "The Big Picture in Addressing Vaccine Hesitancy," *Human Vaccines and Immunotherapeutics* 10, no. 9 (2014): 2600–2602.
47. R. J. Limaye, D. J. Opel, A. Dempsey, M. Ellingson, C. Spina, S. B. Omer, M. Z. Dudley, D. A. Salmon, and S. T. O'Leary, "Communicating with Vaccine-Hesitant Parents: A Narrative Review," *Academic Pediatrics* 21, no. 4 (2021): S24–29.
48. D. Centola, *How Behavior Spreads: The Science of Complex Contagions*, vol. 3 (Princeton University Press, 2018), 14.
49. N. Thacker, "We Are at a Crossroads for Polio Eradication in India. My Experience Shows Why It's Still Achievable," Bill and Melinda Gates Foundation, October 13, 2022. https://www.gatesfoundation.org/ideas/articles/crossroads-for-polio-eradication-my-experience-in-india
50. D. A. Gust, N. Darling, A. Kennedy, and B. Schwartz, "Parents with Doubts About Vaccines: Which Vaccines and Reasons Why," *Pediatrics* 122, no. 4 (2008): 718–25.
51. D. Centola, *Change: How to Make Big Things Happen* (John Murray, 2021).
52. Centola, *How Behavior Spreads*.
53. Although it is not known exactly how large the number of early adopters must be to trigger a tipping point, one study estimated it to be 25 percent of the population. See D. Centola, J. Becker, D. Brackbill, and A. Baronchelli, "Experimental Evidence for Tipping Points in Social Convention," *Science* 360, no. 6393 (2018): 1116–19.
54. Solomon, "Involvement of Civil Society."
55. "Attitudes on Same-Sex Marriage," Pew Research Center, May 14, 2019. https://www.pewresearch.org/religion/fact-sheet/changing-attitudes-on-gay-marriage/
56. Report: "Growing Support for Gay Marriage: Changed Minds and Changing Demographics," Pew Research Center, March 20, 2013. https://www.pewresearch.org/politics/2013/03/20/growing-support-for-gay-marriage-changed-minds-and-changing-demographics/
57. A. Giddens, *The Consequences of Modernity* (Polity, 1990), 85.
58. "Doctors and Scientists Are Seen as the World's Most Trustworthy Professions," *Ipsos*, August 1, 2022. https://www.ipsos.com/en-us/news-polls/global-trustworthiness-index-2022
59. C. Funk, A. Tyson, B. Kennedy, and C. Johnson, "Science and Scientists Held in High Esteem across Global Publics," Pew Research Center, September 29, 2020. https://www.pewresearch.org/science/2020/09/29/science-and-scientists-held-in-high-esteem-across-global-publics/

60. C. Funk, "Mixed Messages about Public Trust in Science," Pew Research Center, December 8, 2017. https://www.pewresearch.org/science/2017/12/08/mixed-messages-about-public-trust-in-science/
61. M. A. Mills, "Why So Many Americans Are Losing Trust in Science," *New York Times*, October 3, 2023. https://www.nytimes.com/2023/10/03/opinion/science-americans-trust-covid.html
62. "Summary Report: Ipsos Survey on Science and the Media," Science Media Centre, released November 23, 2023. https://www.sciencemediacentre.org/wp-content/uploads/2023/11/PDF-summary-report.pdf
63. "How Can Scientists Make the Most of the Public's Trust in Them?," editorial, *Nature* 626 January 31, 2024.
64. J. C. Besley and A. Dudo, *Strategic Science Communication: A Guide to Setting the Right Objectives for More Effective Public Engagement* (Johns Hopkins University Press, 2022).

Chapter 3. From Impact to Encounter

1. J. Friedman-Rudovsky, "Taking the Measure of Madidi," *Science* 337, no. 6092 (2012): 285–87.
2. "Greater Madidi Landscape Conservation Area (Bolivia) The Global Conservation Program: Achievements and Lessons Learned from 10 Years of Support for Threats-Based Conservation at a Landscape and Seascape Scale," Wildlife Conservation Society, La Paz, Bolivia, 2010.
3. Z. Lehm, H. Salas, E. Salinas, I. Gomez, and K. Lara, "Diagnóstico de actores sociales PNANMI—Madidi," Wildlife Conservation Society/CARE, 2002; R. Silva, D. Robison, S. McKean, and P. Álvarez, "La historia de la ocupación del espacio y el uso de los recursos en el PNANMI Madidi y su zona de influencia," Agroecología Bosque y Selva Wildlife Conservation Society/CARE, 2002.
4. A. H. Toomey, "The Making of a Conservation Landscape," *Conservation and Society* 18, no. 1 (2020): 25–36.
5. L. T. Smith, *Decolonizing Methodologies: Research and Indigenous Peoples*, 2nd ed. (Zed Books, 2012), 1.
6. L. E. Alonso, J. L. Deichmann, S. A. McKenna, P. Naskrecki, and S. J. Richards, eds., *Still Counting: Biodiversity Exploration for Conservation the First 20 Years of the Rapid Assessment Program* (Conservation International, 2011).
7. A. Chicchón, "Working with Indigenous Peoples to Conserve Nature: Examples from Latin America," *Conservation and Society* 7, no. 1 (2009): 15–20.
8. T. Whitten, D. Holmes, and K. MacKinnon, "Conservation Biology: A Displacement Behavior for Academia?," *Conservation Biology* 15, no. 1 (2001):

1–3; A. T. Knight, R. M. Cowling, M. Rouget, A. Balmford, A. T. Lombard, and B. M. Campbell, "Knowing but Not Doing: Selecting Priority Conservation Areas and the Research–Implementation Gap," *Conservation Biology* 22, no. 3 (2008): 610–17; D. R. Williams, A. Balmford, and D. S. Wilcove, "The Past and Future Role of Conservation Science in Saving Biodiversity," *Conservation* Letters 13, no. 4 (2020): e12720.

9. M. E. Soulé, "What Is Conservation Biology?," *BioScience* 35, no. 11 (1985): 727–34.
10. D. Ehrenfeld, "War and Peace and Conservation Biology," *Conservation Biology* 14, no. 1 (2000): 106.
11. M. E. Soulé, "Conservation Biology and the 'Real World,'" in *Conservation Biology: The Science of Scarcity and Diversity*, ed. M. E. Soulé (Sinauer Associates, 1986), 5.
12. C. Wilkinson, "Evidencing Impact: A Case Study of UK Academic Perspectives on Evidencing Research Impact," *Studies in Higher Education* 44, no. 1 (2019): 72–85.
13. "Broader Impacts," *National Science Foundation*, accessed April 11, 2024. https://new.nsf.gov/funding/learn/broader-impacts. See also a critique of how this has been operationalized in B. Bozeman and C. Boardman, "Broad Impacts and Narrow Perspectives: Passing the Buck on Science and Social Impacts," *Social Epistemology* 23, no. 3–4 (2009): 183–98.
14. "Special Issue: Impact: The Search for the Science That Matters," *Nature* 502, no. 7471 (2013). Examples of approaches to incentivize nonacademic impact in research include Hong Kong's Research Assessment Exercise, the Swedish Research Council Strategic Research Centre's requirement for impact case studies, and Australia's Engagement and Impact Assessment, among others. See M. S. Reed, M. Ferré, J. Martin-Ortega, R. Blanche, R. Lawford-Rolfe, M. Dallimer, and J. Holden, "Evaluating Impact from Research: A Methodological Framework," *Research Policy* 50, no. 4 (2021): 104147.
15. For an overview of the scholarly debate on how research impact is framed and evaluated across different academic contexts, see, for example, K. E. Smith, J. Bandola-Gill, N. Meer, E. Stewart, and R. Watermeyer, *The Impact Agenda: Controversies, Consequences and Challenges* (Policy Press, 2020).
16. A. H. Toomey, A. T. Knight, and J. Barlow, "Navigating the Space between Research and Implementation in Conservation," *Conservation Letters* 10, no. 5 (2017): 619–25.
17. M. J. Simis, H. Madden, M. A. Cacciatore, and S. K. Yeo, "The Lure of Rationality: Why Does the Deficit Model Persist in Science Communication?" *Public understanding of science* 25, no. 4 (2016): 400-14.

18. For example, see M. C. Nisbet and D. A. Scheufele, "What's Next for Science Communication? Promising Directions and Lingering Distractions," *American Journal of Botany* 96, no. 10 (2009): 1767–78; and D. Sarewitz, "How Science Makes Environmental Controversies Worse," *Environmental Science and Policy* 7, no. 5 (2004): 385–403.
19. R. Pain, "Impact: Striking a Blow or Walking Together?," *ACME: An International E-Journal for Critical Geographies* 13, no. 1 (2014): 19–23; G. Williams, "Researching with Impact in the Global South? Impact-Evaluation Practices and the Reproduction of 'Development Knowledge,'" *Contemporary Social Science* 8, no. 3 (2013): 223–36.
20. See K. A. Kainer, M. L. DiGiano, A. E. Duchelle, L. H. Wadt, E. Bruna, and J. L. Dain, "Partnering for Greater Success: Local Stakeholders and Research in Tropical Biology and Conservation," *Biotropica* 41, no. 5 (2009): 555–62; and R. J. Smith, D. Veríssimo, N. Leader-Williams, R. M. Cowling, and A. T. Knight, "Let the Locals Lead," *Nature* 462 (2009).
21. A. Balmford and R. M. Cowling, "Fusion or Failure? The Future of Conservation Biology," *Conservation Biology* 20, no. 3 (2006): 692.
22. Reed et al., "Evaluating Impact from Research."
23. My research partners in Bolivia included Igor Patzi Sanjinés, María Eugenia Copa Alvaro, and Armando (Mandu) Medinaceli. I also had formal research agreements with the Bolivian chapter of the Wildlife Conservation Society, the Herbario Nacional de Bolivia, and the Tsimane-Mosetén, Takana, and San José de Uchupiamonas Indigenous regional councils.
24. See detailed methods as described in A. H. Toomey, *Who's at the Gap between Research and Implementation? The Places and Spaces of Encounter between Scientists and Local People in Madidi, Bolivia* (Lancaster University [UK], 2015).
25. A. H. Toomey, "What Happens at the Gap between Knowledge and Practice? Spaces of Encounter and Misencounter between Environmental Scientists and Local People," *Ecology and Society* 21, no. 2 (2016).
26. Principal investigators listed on each project were asked to verify the information obtained and queried about the research. Questions asked included: Did the research have any implications for management and/or policy? (What were they?) Were the research results disseminated? (How?) Were the research results published? (Where?) Did the research lead to any management decision or action? See A. H. Toomey, M. E. Copa Alvaro, M. Aiello-Lammens, O. Loayza Cossio, and J. Barlow, "A Question of Dissemination: Assessing the Practices and Implications of Research in Tropical Landscapes," *Ambio* 48 (2019): 35–47.
27. Toomey et al., "A Question of Dissemination." See also A. H. Toomey,

"Redefining 'Impact' So Research Can Help Real People Right Away, Even before Becoming a Journal Article," *The Conversation*, May 7, 2018. https://theconversation.com/redefining-impact-so-research-can-help-real-people-right-away-even-before-becoming-a-journal-article-94219

28. In interviews, researched provided many reasons for this lack of dissemination. This will be further explored in chapter 7. See also Toomey, "Who's at the Gap?"
29. N. C. Pitman, M. Del C. Azáldegui, K. Salas, G. Vigo, and D. A. Lutz, "Written Accounts of an Amazonian Landscape over the Last 450 Years," *Conservation Biology* 21, no. 1 (2007): 253–62.
30. T. Tarifa, "Desarrollo y perspectivas de la Mastozoología en Bolivia: Una historia de pioneros bolivianos y padres extranjeros," *Mastozoología Neotropical* 12, no. 2 (2005): 125–32.
31. Interview with Celín Quenevo, July 2014. See also "Indigenous Perspectives on Research," released July 6, 2015 on YouTube (https://www.youtube.com/watch?v=uQd95Nq05Rk).
32. Workshop with botanists from the Herbario Nacional de Bolivia, August 2013.
33. The first "self-monitoring" research project in the region (which served as a model for the Wildlife Conservation Society project) was carried out in the neighboring Pilón Lajas, led by biologist Wendy Townsend and funded by Veternarios Sin Fronteras. See also M. E. Copa and W. R. Townsend, "Aprovechamiento De La Fauna Por Dos Comunidades Tsimane: Un Subsidio Del Bosque a La Economia Familiar," *Revista Boliviana de Ecología y Conservacion Ambiental* 16 (2004): 41–48.
34. Toomey, "Making of a Conservation Landscape."
35. María Copa, Zoom interview, December 18, 2023.
36. Workshop with park guards from Pilón Lajas, November 2013.
37. Eduardo Cavinas, community member of Cachichira, semistructured interview, September 2012.
38. Copa, interview.
39. A. H. Toomey and M. Copa, "Informe Preliminar: Valoración De Actividades De Monitoreo Por Pobladores Locales," report submitted to Wildlife Conservation Society Bolivia (2013).
40. G. W. Allport, *The Nature of Prejudice* (Addison-Wesley, 1954), 263.
41. For other scholars who have built on the ideas of the contact hypothesis, see work by María Torre: M. E. Torre, "The History and Enactments of Contact in Social Psychology" (PhD diss., City University of New York, 2010); and M. E. Torre, M. Fine, N. Alexander, A. B. Billups, Y. Blanding, E. Genao, E.

Marboe, T. Salah, and K. Urdang, "Participatory Action Research in the Contact Zone," in *Youth Participatory Action Research in Motion*, ed. J. Cammarota and M. Fine (Routledge, 2008). See also M. L. Pratt, *Imperial Eyes: Travel Writing and Transculturation*. (Routledge, 1992).
42. Toomey, "What Happens at the Gap."
43. D. J. Roux, K. H. Rogers, H. C. Biggs, P. J. Ashton, and A. Sergeant, "Bridging the Science–Management Divide: Moving from Unidirectional Knowledge Transfer to Knowledge Interfacing and Sharing," *Ecology and Society* 11, no. 1 (2006).
44. M. K. Halpern and K. C. Elliott, "Science as Experience: A Deweyan Model of Science Communication," *Perspectives on Science* 30, no. 4 (2022): 621–56.
45. Some terms used to describe research that involves non-academic partners, communities, policymakers and other interested, impacted, or relevant groups throughout the entire process include *participatory action research, coproduced research, community-led research*, and *use-inspired research*. For a list of terms and definitions, see L. M. Vaughn and F. Jacquez, "Participatory Research Methods—Choice Points in the Research Process," *Journal of Participatory Research Methods* 1, no. 1 (2020).
46. Some key thinkers are Paulo Freire (critical pedagogy), Orlando Fals Borda (sociology), Kurt Lewin (social psychology), and Bagele Chilisa (Indigenous methodologies). See also A. McIntyre, *Participatory Action Research* (Sage, 2007).
47. As decades of development and aid scholarship have shown, participation fatigue is a very real phenomenon. B. Cooke and U. Kothari, *Participation: The New Tyranny?* (Zed Books, 2001).
48. Vaughn and Jacquez, "Participatory Research Methods." See also J. L. Shirk, H. L. Ballard, C. C. Wilderman, T. Phillips, A. Wiggins, R. Jordan, E. McCallie, et al., "Public Participation in Scientific Research: A Framework for Intentional Design," *Ecology and Society* 17, no. 2 (2012).

Chapter 4. Asking a Good Question

1. At the time of the interview, Molly Cheatum was the watershed restoration program manager for the Chesapeake Bay Foundation (CBF). The goal of the project was to create natural buffers that reduced nutrient runoff into waterways and protected local aquatic life, and she had obtained a grant to plant trees on farms across the region. Bill Chain was the agricultural program manager for CBF. Both Cheatum and Chain have since left CBF and now run a company (Swift Aeroseed LLC) that uses drones to seed cover crops.
2. H. Blanco-Canqui, M. M. Mikha, J. G. Benjamin, L. R. Stone, A.J. Schlegel,

D. J. Lyon, M. F. Vigil, and P. W. Stahlman, "Regional Study of No-Till Impacts on Near-Surface Aggregate Properties That Influence Soil Erodibility," *Soil Science Society of America Journal* 73, no. 4 (2009): 1361–68; R. A. Peiretti, "No Till Improves Soil Functioning and Water Economy," Food and Agriculture Organization of the United Nations, retrieved October 23, 2020.
3. Dave McLaughlin, interview, June 2021.
4. McLaughlin, interview.
5. J. R. Anderson and G. Feder, "Agricultural Extension," *Handbook of Agricultural Economics* 3 (2007): 2343–78; E. S. Gornish, and L. M. Roche, "Cooperative Extension Is Key to Unlocking Public Engagement with Science," *Frontiers in Ecology and the Environment* 15, no. 9 (2017): 487–88.
6. John Tooker, Zoom interview, August 24, 2023.
7. Tooker, interview.
8. Maggie Douglas, interview, August 22, 2023.
9. M. R. Douglas and J. F. Tooker, "Slug (Mollusca: Agriolimacidae, Arionidae) Ecology and Management in No-Till Field Crops, with an Emphasis on the Mid-Atlantic Region," *Journal of Integrated Pest Management* 3, no. 1 (2012): C1–9.
10. D. Klingelhöfer, M. Braun, D. Brüggmann, and D. A. Groneberg, "Neonicotinoids: A Critical Assessment of the Global Research Landscape of the Most Extensively Used Insecticide," *Environmental Research* 213 (2022): 113727.
11. D. A. Stanley, M. P. Garratt, J. B. Wickens, V. J. Wickens, S. G. Potts, and N. E. Raine, "Neonicotinoid Pesticide Exposure Impairs Crop Pollination Services Provided by Bumblebees," *Nature* 528, no. 7583 (2015): 548–50; J. Hopwood, M. Vaughan, M. Shepherd, D. Biddinger, E. Mader, S. H. Black, and C. Mazzacano, "Are Neonicotinoids Killing Bees? A Review of Research into the Effects of Neonicotinoid Insecticides on Bees, with Recommendations for Action," *Xerces Society for Invertebrate Conservation, USA* (2012).
12. Neonicotinoids do not accumulate in the slugs, but rather travel through the slugs to affect the beetles. Because neonicotinoids are water soluble, they behave somewhat differently in food webs than fat-soluble products like DDT that bioaccumulate. Maggie Douglas, personal communication, December 22, 2023.
13. The trial was conducted on the 1,500-acre farm of Lucas Criswell, who grows corn, soybeans, wheat, rye, and canola in Lewisburg, Pennsylvania, using no-till practices.
14. 67 percent more slugs, 31 percent fewer predators, 19 percent fewer soybean plants, and 5 percent lower yield. M. R. Douglas, J. R. Rohr, and J. F. Tooker, "Editor's Choice: Neonicotinoid Insecticide Travels through a Soil Food

Chain, Disrupting Biological Control of Non-Target Pests and Decreasing Soya Bean Yield," *Journal of Applied Ecology* 52, no. 1 (2015): 250–60.
15. Tooker, interview.
16. There was also evidence coming out of other labs that suggested that neonicotinoid-treated seeds have little to no yield benefit under typical conditions. See, for example, A. de Freitas Bueno, M. J. Batistela, R. C. Oliveira de Freitas Bueno, J. de Barros França-Neto, M. A. Naime Nishikawa, and A. Libério Filho, "Effects of Integrated Pest Management, Biological Control and Prophylactic Use of Insecticides on the Management and Sustainability of Soybean," *Crop Protection* 30, no. 7 (2011): 937–45; and M. P. Seagraves and J. G. Lundgren, "Effects of Neonicitinoid Seed Treatments on Soybean Aphid and Its Natural Enemies," *Journal of Pest Science* 85 (2012): 125–32.
17. J. Kleinschmit and B. Lilliston, "Unknown Benefits, Hidden Costs," *Institute for Agriculture and Trade Policy* (2015).
18. The visit was facilitated by the Center for Food Safety and prompted in part by a report released in 2014 by the US Environmental Protection Agency that concluded that neonicotinoids provide negligible overall benefits to soybean production in most situations. The congressional briefing was organized by US Representatives John Conyers and Earl Blumenauer. John Tooker, personal communication, March 2024.
19. Tooker, interview.
20. Also important to consider is how industry-funded science has promoted neonicotinoids (and other product-based pest management approaches) to farmers by emphasizing the benefits while diminishing the risks or downsides of treatments. This practice can lead policy makers to be reluctant to ban products that are seen by farmers as essential to maintaining productivity and their economic bottom line. See J. F. Tooker, M. R. Douglas, and C. H. Krupke, "Neonicotinoid Seed Treatments: Limitations and Compatibility with Integrated Pest Management," *Agricultural and Environmental Letters* 2, no. 1 (2017): ael2017.08.0026.
21. United States Department of Agriculture, "E595F: Improving Soil Organism Habitat on Agricultural Land," USDA Conservation Enhancement Activity, April 2021. https://www.nrcs.usda.gov/sites/default/files/2022-11/E595F%20April%202021.pdf
22. Douglas, interview.
23. Douglas, Rohr, and Tooker, "Editor's Choice."
24. M. P. Hill, S. Macfadyen, and M. A. Nash, "Broad Spectrum Pesticide Application Alters Natural Enemy Communities and May Facilitate Secondary Pest Outbreaks," *PeerJ* 5 (2017): e4179.

25. Tooker, interview.
26. McLaughlin, interview.
27. Personal communications with Dave McLaughlin and Lucas Criswell, February and March 2024. Lucas Criswell also said that although nontreated seed is available for corn crops, those seeds are not yet as good quality as the treated varieties.
28. National Caucus of Environmental Legislators, "New York Enacts Nation's First Neonicotinoid Treated Seed Ban," news release, January 3, 2024. https://www.ncelenviro.org/articles/new-york-enacts-nations-first-neonicotinoid-treated-seed-ban/#:~:text=On%20December%202022%2C%20New%20York,soybean%2C%20and%20wheat%20agricultural%20production Two other states that border Pennsylvania, Maryland and New Jersey, have also enacted laws that restrict neonicotinoid use, although not yet in seeds for commercial agriculture. "States Make Way for Pesticide Reforms," Xerces Society for Invertebrate Conservation, July 27, 2023. https://xerces.org/blog/states-make-way-for-pesticide-reforms
29. Bill Chain, personal communication, June 2021.
30. Dave McLaughlin, phone conversation, January 9, 2024. See also "Pennsylvania No Till Alliance." https://panotill.org/
31. R. W. Kimmerer, *Braiding Sweetgrass: Indigenous Wisdom, Scientific Knowledge, and the Teachings of Plants* (Milkweed Editions, 2013), 60. Kimmerer's book emphasizes the value of Indigenous knowledge for understanding, acknowledging, and celebrating the living world.
32. V. Bush, *Science—The Endless Frontier: A Report to the President on a Program for Postwar Scientific Research* (Office of Scientific Research Development, 1945).
33. Between 1953 and 2013, federal funding for "basic" research allocated to colleges and universities increased from $82 million to more than $22 billion annually, while during the same time period, the amount allocated for "applied" research increased from $59 million to less than $10 billion. See "U.S. Applied Research Expenditures, by Source of Funds and Performing Sector: 1952–2013," National Science Board—Science and Engineering Indicators 2016, accessed April 11, 2024 (https://www.nsf.gov/statistics/2016/nsb20161/uploads/1/7/at04-08.pdf); and "U.S. Basic Research Expenditures, by Source of Funds and Performing Sector: 1952–2013," National Science Board—Science and Engineering Indicators 2016, accessed April 11, 2024 (https://www.nsf.gov/statistics/2016/nsb20161/uploads/1/7/at04-03.pdf).
34. "The Golden Goose Award." https://www.goldengooseaward.org/
35. D. Sarewitz, "Saving Science," *New Atlantis*, Spring/Summer 2016. https://www.thenewatlantis.com/publications/saving-science

36. D. Heady, "25 Years of Herceptin: A Groundbreaking Advancement in Breast Cancer Treatment," *UCLA Health* (October 2023). https://www.uclahealth.org/news/25-years-herceptin-groundbreaking-advancement-breast-cancer
37. Sarewitz, "Saving Science."
38. D. van Miert, ed., *Communicating Observations in Early Modern Letters (1500–1675): Epistolography and Epistemology in the Age of the Scientific Revolution* (Warburg Institute, 2013).
39. T. S. Kuhn, *The Structure of Scientific Revolutions* (University of Chicago Press, 1962).
40. S. A. Sloman and P. Fernbach, *The Knowledge Illusion: Why We Never Think Alone* (Penguin, 2018). See also R. I. M. Dunbar, "Neocortex Size as a Constraint on Group Size in Primates," *Journal of Human Evolution* 22, no. 6 (1992): 469–93.
41. T. W. Malone, *Superminds: The Surprising Power of People and Computers Thinking Together* (Little, Brown Spark, 2018); H. Mercier and D. Sperber, *The Enigma of Reason.* (Harvard University Press, 2017); B. Maciejovsky and D. V. Budescu, "Collective Induction without Cooperation? Learning and Knowledge Transfer in Cooperative Groups and Competitive Auctions," *Journal of Personality and Social Psychology* 92, no. 5 (2007): 854.
42. P. C. Wason, "Reasoning about a Rule," *Quarterly Journal of Experimental Psychology* 20, no. 3 (1968): 273–81; E. M. Roth, "Facilitating Insight in a Reasoning Task," *British Journal of Psychology* 70, no. 2 (1979): 265–71.
43. D. Moshman and M. Geil, "Collaborative Reasoning: Evidence for Collective Rationality," *Thinking and Reasoning* 4, no. 3 (1998): 231–48.
44. G. Henriques, "Groupthink and the Evolution of Reason Giving," in *Groupthink in Science: Greed, Pathological Altruism, Ideology, Competition, and Culture*, eds. D. M. Allen and J. W. Howell (Springer, 2020): 15–25.
45. A. L. Antonio, M. J Chang, K. Hakuta, D. A. Kenny, S. Levin, and J. F. Milem, "Effects of Racial Diversity on Complex Thinking in College Students," *Psychological science* 15, no. 8 (2004): 507–10.
46. D. M. Allen and J. W. Howell, eds., *Groupthink in Science: Greed, Pathological Altruism, Ideology, Competition, and Culture* (Springer, 2020).
47. N. Oreskes, *Why Trust Science?* (Princeton University Press, 2019).
48. S. Begley, "The Maddening Saga of How an Alzheimer's 'Cabal' Thwarted Progress toward a Cure for Decades," *STAT* (2019). https://www.statnews.com/2019/06/25/alzheimers-cabal-thwarted-progress-toward-cure/
49. C. Behl, *Alzheimer's Disease Research: What Has Guided Research So Far and Why It Is High Time for a Paradigm Shift* (Springer, 2023).
50. N. S. Smith, I. M. Côté, L. Martinez-Estevez, E. J. Hind-Ozan, A. L. Quiros,

N. Johnson, S. J. Green, et al., "Diversity and Inclusion in Conservation: A Proposal for a Marine Diversity Network," *Frontiers in Marine Science* 4 (2017): 234. See also A. W. Woolley, C. F. Chabris, A. Pentland, N. Hashmi, and T. W. Malone, "Evidence for a Collective Intelligence Factor in the Performance of Human Groups," *Science* 330, no. 6004 (2010): 686–88.

51. T. H. Swartz, A. S. Palermo, S. K. Masur, and J. A. Aberg, "The Science and Value of Diversity: Closing the Gaps in Our Understanding of Inclusion and Diversity," *Journal of Infectious Diseases* 220, no. suppl._2 (2019): S33–41.

52. K. Rolin, "The Epistemic Significance of Diversity," in *The Routledge Handbook of Social Epistemology*, eds. M. Fricker, P. J. Graham, D. Henderson, and N. J. Pedersen (Routledge, 2019), 158–66.

53. Key here is language of "transdisciplinarity," which is often conflated with interdisciplinarity. Transdisciplinary refers to research that goes beyond academia and thus requires the inclusion of perspectives from nonscientists. A. H. Toomey, N. Markusson, E. Adams, and B. Brockett, "Inter- and Trans-Disciplinary Research: A Critical Perspective," *GSDR Brief* (2015): 1–3.

54. Matt Bomgardner, interview, June 2021.

55. Bomgardner, interview.

56. P. B. Stark, "Nullius in Verba," in *The Practice of Reproducible Research: Case Studies and Lessons from the Data-Intensive Sciences*, eds. J. Kitzes, D. Turek, and F. Deniz (University of California Press, 2017).

57. S. Shapin and S. Schaffer, *Leviathan and the Air-Pump: Hobbes, Boyle, and the Experimental Life* (Princeton University Press, 2011).

58. J. Fairhead and M. Leach, *Misreading the African Landscape: Society and Ecology in a Forest-Savanna Mosaic* (Cambridge University Press, 1996).

59. This point is particularly true in terms of Indigenous or "traditional" knowledges, as scholars and Indigenous peoples have raised concerns about the tendency of researchers to find ways to make cultural beliefs and practices conform to western conceptions of knowledge, thus ignoring power relations and resulting in the creation of information that serves scientists rather than those from whom such knowledge comes. See E. L. Bohensky and Y. Maru, "Indigenous Knowledge, Science, and Resilience: What Have We Learned from a Decade of International Literature on 'Integration'?," *Ecology and Society* 16, no. 4 (2011). See also C. Hird, D. M. David-Chavez, S. Spang Gion, and V. van Uitregt, "Moving Beyond Ontological (Worldview) Supremacy: Indigenous Insights and a Recovery Guide for Settler-Colonial Scientists," *Journal of Experimental Biology* 226, no. 12 (2023).

60. "Grants Available from Northeast SARE." https://www.sare.org/wp-content/uploads/Northeast_SARE_Grant_Comparison_09102021.png

61. Candice Huber, personal communication, April 2, 2024: From past calls for proposals, "an advisory committee consisting of at least two farmers and one agricultural service provider (e.g., Extension, NRCS or other federal or state agency, private or non-profit organizations, veterinarian or other farm advisor) is required to provide input on project design and implementation. Additional people representing other affiliations, such as researchers, may also be included on the committee."
62. S. Malone and J. Weiss-Wolf, "America Lost Its Way on Menopause Research. It's Time to Get Back on Track," *Washington Post*, April 28, 2022. https://www.washingtonpost.com/opinions/2022/04/28/menopause-hormone-therapy-nih-went-wrong/
63. A. A. Mirin, "Gender Disparity in the Funding of Diseases by the US National Institutes of Health," *Journal of Women's Health* 30, no. 7 (2021): 956–63.
64. F. Farooq, P. J. Mogayzel, S. Lanzkron, C. Haywood, and J. J. Strouse, "Comparison of Us Federal and Foundation Funding of Research for Sickle Cell Disease and Cystic Fibrosis and Factors Associated with Research Productivity," *JAMA Network Open* 3, no. 3 (2020): e201737–e37.
65. Douglas, interview.
66. Douglas, interview.
67. M. H. O'Brien, "Being a Scientist Means Taking Sides," *BioScience* 43, no. 10 (1993): 707.
68. Tooker, interview.

Chapter 5. The Privilege of Choice: Methods, Permissions, and Location

1. T. L. Beauchamp, "The Belmont Report," in *The Oxford Textbook of Clinical Research Ethics*, eds. E. J. Emanuel et al. (Oxford University Press, 2011).
2. For a definition and overview of parachute science, see B. Odeny, and R. Bosurgi, "Time to End Parachute Science," *PLOS Medicine* 19, no. 9 (2022): e1004099.
3. M. L. Pratt, *Imperial Eyes: Travel Writing and Transculturation* (London and New York: Routledge, 1992). (Excerpt from William Paterson, p. 56 in Pratt.)
4. S. Harding, "After the Neutrality Ideal: Science, Politics, and 'Strong Objectivity,'" *Social Research* 59, no. 3 (1992): 567–87.
5. D. B. Paul and H. G. Spencer, "The Hidden Science of Eugenics," *Nature* 374, no. 6520 (1995): 302–4.
6. C. H. Dupree and C. M. Boykin, "Racial Inequality in Academia: Systemic Origins, Modern Challenges, and Policy Recommendations," *Policy Insights from the Behavioral and Brain Sciences* 8, no. 1 (2021): 11–18.

7. G. Vogel, "Natural History Museums Face Their Own Past," *Science* 363, no. 6434 (March 2019): 1371–72.
8. P. V. Stefanoudis, W. Y. Licuanan, T. H. Morrison, S. Talma, J. Veitayaki, and L. C. Woodall, "Turning the Tide of Parachute Science," *Current Biology* 31, no. 4 (2021): R184–85.
9. "Is Science Only for the Rich?," *Nature* 537, no. 7621 (2016): 466–70.
10. Odeny and Bosurgi, "Time to End Parachute Science."
11. P. V. Stefanoudis, W. Y. Licuanan, T. H. Morrison, S. Talma, J. Veitayaki, and L. C. Woodall, "Turning the Tide of Parachute Science," *Current Biology* 31, no. 4 (2021): R184–85.
12. R. O. Ajwang' and L. Edmondson, "Love in the Time of Dissertations: An Ethnographic Tale," *Qualitative Inquiry* 9, no. 3 (2003): 466–80.
13. "Too Many Academics Study the Same People," editorial, *Nature* 551, no. 7679 (2017): 141–42.
14. J. Koen, D. Wassenaar, and N. Mamotte, "The 'Over-Researched Community': An Ethics Analysis of Stakeholder Views at Two South African Hiv Prevention Research Sites," *Social Science and Medicine* 194 (2017): 1–9.
15. C. Button and G. T. Aiken, *Over Researched Places: Towards a Critical and Reflexive Approach* (Taylor and Francis, 2022).
16. H. A. Ruszczyk, "Overlooked Cities and Under-Researched Bharatpur, Nepal," in C. Button and G. T. Aiken, *Over Researched Places: Towards a Critical and Reflexive Approach* (Taylor and Francis, 2022), 104.
17. Ruszczyk, "Overlooked Cities," 102.
18. M. Mcdonnel, "The History of Urban Ecology: An Ecologist Perspective" in *Urban Ecology: Patterns, Processes and Applications*, ed. J Niemala (Oxford Academic, 2011), 5–13; A. R. Berkowitz, C. H. Nilon, K. S. Hollweg, and M. V. Melosi, "The Historical Dimension of Urban Ecology: Frameworks and Concepts," in *Understanding Urban Ecosystems: A New Frontier for Science and Education* (Springer, 2003): 187–200.
19. J. P. Collins, A. Kinzig, N. B. Grimm, W. F. Fagan, D. Hope, J. Wu, and E. T. Borer, "A New Urban Ecology: Modeling Human Communities as Integral Parts of Ecosystems Poses Special Problems for the Development and Testing of Ecological Theory," *American Scientist* 88, no. 5 (2000): 416–25.
20. E. Plänitz, "Neglecting the Urban? Exploring Rural-Urban Disparities in the Climate Change–Conflict Literature on Sub-Sahara Africa," *Urban Climate* 30 (2019): 100533.
21. M. P. Andrasik, G. B. Broder, S. E. Wallace, R. Chaturvedi, N. L. Michael, S. Bock, C. Beyrer, et al. "Increasing Black, Indigenous and People of Color

Participation in Clinical Trials through Community Engagement and Recruitment Goal Establishment," *PLOS ONE* 16, no. 10 (2021): e0258858.

22. P. Louie and R. Wilkes, "Representations of Race and Skin Tone in Medical Textbook Imagery," *Social Science and Medicine* 202 (2018): 38–42.

23. D. P. Ly, "Black-White Differences in the Clinical Manifestations and Timing of Initial Lyme Disease Diagnoses," *Journal of General Internal Medicine* 37, no. 10 (2022): 2597–600.

24. A. Oza, E. Kwong, T. Lu, G. Spitzer, "COVID-19 Made Pulse Oximeters Ubiquitous. Engineers Are Fixing Their Racial Bias," National Public Radio, February 23, 2023. https://www.npr.org/2023/02/10/1156166554/covid-19-pulse-oximeters-racial-bias; For a detailed exploration of how scientists have long neglected important biophysical questions on melanin, see chapter 5, "The Physics of Melanin," in C. Prescod-Weinstein, *The Disordered Cosmos: A Journey into Dark Matter, Spacetime, and Dreams Deferred* (Hachette UK, 2021).

25. J. Opara, "Long-Term Funding Needed to Nurture African Science," *SciDev.Net*, April 17, 2023. https://www.gavi.org/vaccineswork/long-term-funding-needed-nurture-african-science

26. L. T. Benuto, J. Singer, J. Casas, F. González, and A. Ruork, "The Evolving Definition of Cultural Competency: A Mixed Methods Study," *International Journal of Psychology and Psychological Therapy* 18, no. 3 (2018): 371–84. Other terms commonly used are *cultural humility* and *cultural safety*.

27. E. Tuck, "Collaborative Indigenous Research Is a Way to Repair the Legacy of Harmful Research Practices," *The Conversation*, November 21, 2022. https://theconversation.com/collaborative-indigenous-research-is-a-way-to-repair-the-legacy-of-harmful-research-practices-193912

28. For example, 17 percent of New Zealand's (Aotearoa's) population identifies as Māori, but Māori scholars comprise only 5 percent of faculty in New Zealand (Aotearoa) universities. A. Movono, A. Carr, E. Hughes, F. Higgens-Desbiolles, J. Hapeta, R. Scheyvens, and R. Stewart-Withers, "Indigenous Scholars Struggle to Be Heard in the Mainstream. Here's How Journal Editors and Reviewers Can Help," *The Conversation*, April 11, 2021 (https://theconversation.com/indigenous-scholars-struggle-to-be-heard-in-the-mainstream-heres-how-journal-editors-and-reviewers-can-help-157860); see also L. T. Smith, E. Tuck, and K. W. Yang, *Indigenous and Decolonizing Studies in Education: Mapping the Long View* (Routledge, 2018).

29. A. Harmon, "Indian Tribe Wins Fight to Limit Research of Its DNA," *New York Times*, April 21, 2010. https://www.nytimes.com/2010/04/22/us/22dna.html

30. S. Moodie, "Power, Rights, Respect and Data Ownership in Academic Research with Indigenous Peoples," *Environmental Research* 110, no. 8 (2010): 818.
31. My master's research (2006–2008) was funded by a US Environmental Protection Agency STAR Research Fellowship ($75,000). My PhD research (2011–2015) was funded by Lancaster University Faculty of Science and Technology PhD Studentship ($81,000).
32. H. Raffles, *In Amazonia: A Natural History* (Princeton University Press, 2002), 177.
33. This ethical issue is both complicated and understudied in research. See T. Molony and D. Hammett, "The Friendly Financier: Talking Money with the Silenced Assistant," *Human Organization* 66, no. 3 (2007): 292–300.
34. A. Greer and J. Buxton, *A Guide for Paying Peer Research Assistants* (City of Vancouver, 2016).
35. Greer and Buxton, *A Guide for Paying*.
36. A. M. Brandt, "Racism and Research: The Case of the Tuskegee Syphilis Study," *Hastings Center Report* 8, no. 6 (1978): 21–29.
37. U. Beck, *Risk Society: Towards a New Modernity* (Sage, 1992).
38. See, for example, A. Irwin, *Citizen Science: A Study of People, Expertise, and Sustainable Development* (Routledge, 1995); S. Jasanoff, "Technologies of Humility: Citizen Participation in Governing Science," *Minerva* 41, no. 3 (2003): 223–44; and H. Nowotny, M. Gibbons, and P. Scott, *Re-Thinking Science: Knowledge and the Public in an Age of Uncertainty* (Polity, 2006).
39. Visit CLEAR's website at https://civiclaboratory.nl/. The About page reads: "Civic Laboratory for Environmental Action Research (CLEAR) is an interdisciplinary natural and social science lab space dedicated to good land relations directed by Dr. Max Liboiron at Memorial University, Canada. Equal parts research space, methods incubator, and social collective, CLEAR's ways of doing things, from environmental monitoring of plastic pollution to how we run lab meetings, are based on values of humility, accountability, and anti-colonial research relations. We specialize in community-based and citizen science monitoring of plastic pollution, particularly in wild food webs, and the creation and use of anti-colonial research methodologies."
40. M. Liboiron, *Pollution Is Colonialism* (Duke University Press, 2021), 139.
41. M. Liboiron, A. Zahara, and I. Schoot, "Community Peer Review: A Method to Bring Consent and Self-Determination into the Sciences," *Preprints*, June 7, 2018: 2018060104.
42. Max Liboiron, Zoom interview, January 4, 2024.
43. Liboiron, *Pollution Is Colonialism*, 142. Liboiron is also quoting from E. Tuck

and M. McKenzie, *Place in Research: Theory, Methodology, and Methods* (Routledge, 2014).
44. J. Brumbaugh, B. A. Aguado, T. Lysaght, and L. S. Goldstein, "Human Fetal Tissue Is Critical for Biomedical Research," *Stem Cell Reports* 18, no. 12 (2023): 2300–2312.
45. "Science Must Right a Historical Wrong," editorial, *Nature* 585, no. 7 (2020).
46. B. Panikkar, N. Smith, and P. Brown, "Reflexive Research Ethics in Fetal Tissue Xenotransplantation Research," *Accountability in Research* 19, no. 6 (2012): 344–69.
47. H.-T. Nguyen, H. Thach, E. Roy, K. Huynh, and C. M.-T. Perrault, "Low-Cost, Accessible Fabrication Methods for Microfluidics Research in Low-Resource Settings," *Micromachines* 9, no. 2 (2018): 461.
48. J. S. Cybulski, J. Clements, and M. Prakash, "Foldscope: Origami-Based Paper Microscope," *PLOS ONE* 9, no. 6 (2014): e98781.
49. "Our Story," Foldscope Instruments, accessed March 28, 2024. https://foldscope.com/pages/our-story
50. M. Ganesan, J. R. Christyraj, S. Venkatachalam, B. V. Yesudhason, K. S. Chelladurai, M. Mohan, K. Kalimuthu, Y. B. Narkhede, J. Durairaj, and J. D. S. Christyraj, "Foldscope Microscope, an Inexpensive Alternative Tool to Conventional Microscopy—Applications in Research and Education: A Review," *Microscopy Research and Technique* 85, no. 11 (2022): 3484–94.
51. M. Minkler, V. B. Vasquez, and P. Shepard, "Promoting Environmental Health Policy through Community Based Participatory Research: A Case Study from Harlem, New York," *Journal of Urban Health* 83 (2006): 101–10.
52. K. H. Jung, K. Bernabé, K. Moors, B. Yan, S. N. Chillrud, R. Whyatt, D. Camann, et al., "Effects of Floor Level and Building Type on Residential Levels of Outdoor and Indoor Polycyclic Aromatic Hydrocarbons, Black Carbon, and Particulate Matter in New York City," *Atmosphere* 2, no. 2 (2011): 96–109.
53. Minkler, Vasquez, and Shepard, "Promoting Environmental Health Policy," 104.
54. "Dirty Diesel Campaign," WE ACT, accessed March 28, 2024. https://www.weact.org/campaigns/dirty-diesel-campaign/
55. S. Epstein, "The Construction of Lay Expertise: Aids Activism and the Forging of Credibility in the Reform of Clinical Trials," *Science, Technology, and Human Values* 20, no. 4 (1995): 408–37.
56. Epstein, "The Construction of Lay Expertise," 421. See also (cited in Epstein) H. Edgar and D. J. Rothman, "New Rules for New Drugs: The Challenge of AIDS to the Regulatory Process," in *AIDS: Society, Ethics and Law* (Routledge, 2018), 447–78; and A. Feenberg, "On Being a Human Subject: Interest and

Obligation in the Experimental Treatment of Incurable Disease," in *Philosophical Forum* 23, no. 3 (1992): 213–30.
57. F. Kearns, *Getting to the Heart of Science Communication: A Guide to Effective Engagement* (Island Press, 2021), 145.
58. M. Waruru, "Renowned Journal Rejects Papers That Exclude African Researchers," *University World News*, June 3, 2022. https://www.universityworldnews.com/post.php?story=20220603115640789
59. D. Haelewaters, T. A. Hofmann, and A. L. Romero-Olivares, "Ten Simple Rules for Global North Researchers to Stop Perpetuating Helicopter Research in the Global South," *PLOS Computational Biology* 17, no. 8 (2021): e1009277.
60. For example, in the Society of Conservation Biology's code of ethics (developed in 2004), point 11 reads: "When working professionally, especially outside their region of residence, interact and collaborate with counterparts, present seminars, confer regularly with appropriate officials, share information, involve colleagues and students in professional activities, contribute to local capacity-building, and equitably share the benefits arising from the use of local knowledge, practices, and genetic resources." Available at https://conbio.org/about-scb/who-we-are/code-of-ethics#:~:text=Protect%20the%20rights%20and%20welfare,sustaining%20natural%20populations%20and%20ecosystems
61. For additional guidance on establishing collaborative research relationships when working abroad, see A. Chin, L. Baje, T. Donaldson, K. Gerhardt, R. W. Jabado, P. M. Kyne, R. Mana, et al., "The Scientist Abroad: Maximising Research Impact and Effectiveness When Working as a Visiting Scientist," *Biological Conservation* 238 (2019): 108231.
62. P. Kalmus, "A Climate Scientist Who Decided Not to Fly," *Grist*, February 21, 2016. https://grist.org/climate-energy/a-climate-scientist-who-decided-not-to-fly/
63. Liboiron, interview.

Chapter 6. The Power of Participation: Data Collection and Analysis

1. B. Alberts, "Restoring Science to Science Education," *Issues in Science and Technology* 25, no. 4 (Summer 2009). https://issues.org/alberts-2/
2. J. DeWitt, L. Archer, and A. Mau, "Dimensions of Science Capital: Exploring Its Potential for Understanding Students' Science Participation," *International Journal of Science Education* 38, no. 16 (2016): 2431–49. For example, one study of more than forty thousand young people in the United Kingdom found that although interest in science was high, this interest did not translate

into aspirations to be a scientist. See ASPIRES, "Supporting Increased and Wider Participation in STEM," UCL Institute of Education, London, 2020.
3. E. Seymour and A. Hunter. *Talking about Leaving Revisited: Persistence, Relocation, and Loss in Undergraduate STEM Education* (Springer, 2019). Although popular among professors (I've given hundreds of lectures over my career), decades of science education research have consistently found that the least effective ways to teach science are through lecture-style teaching and the use of textbooks. Lecture-style teaching tends to engage very low levels of cognitive function and has been found to result in the lowest knowledge retention rate of any method of learning. See J. T. DiPiro, "Why Do We Still Lecture?," *American Journal of Pharmaceutical Education* 73, no. 8 (2009).
4. Seymour and Hunter, *Talking about Leaving Revisited*. For example, research has shown that Black students in the United States have similar rates of intention to pursue STEM degrees as compared to their White peers but are more likely to switch majors early in their academic career due to negative experiences, including subtle or overt forms of racial bias from teachers, administrators, and other students. See M. Gasman and T. Nguyen, *Making Black Scientists: A Call to Action* (Harvard University Press, 2019); and C. Prescod-Weinstein, *The Disordered Cosmos: A Journey into Dark Matter, Spacetime, and Dreams Deferred* (Hachette UK, 2021). Similar patterns are experienced by Indigenous students across Latin America and Africa, by itinerant groups in Europe, and by ethnic minorities in China and India. Regardless of the context, research has found that the problem of engagement is not due to lack of interest but rather to a direct result of limited access, support, and guidance. See also B. Wong, *Science Education, Career Aspirations and Minority Ethnic Students* (Springer, 2016).
5. L. Archer, E. Dawson, J. DeWitt, A. Seakins, and B. Wong, "'Science Capital': A Conceptual, Methodological, and Empirical Argument for Extending Bourdieusian Notions of Capital Beyond the Arts," *Journal of Research in Science Teaching* 52, no. 7 (2015): 922–48.
6. L. Archer and J. DeWitt, *Understanding Young People's Science Aspirations* (Routledge, 2016).
7. STEM Learning, "Science Capital: Making Science Relevant," July 16, 2019. https://www.stem.org.uk/news-and-views/opinions/science-capital-making-science-relevant
8. M. K. Halpern, and K. C. Elliott, "Science as Experience: A Deweyan Model of Science Communication," *Perspectives on Science* 30, no. 4 (2022): 621–56.
9. See works of John Dewey, Kurt Lewin, Kurt Hahn, and Jean Piaget.
10. A. Prashant, D. Devegowda, P. Vishwanath, and S. M. Nataraj, "Impact of

Experiential Learning among Medical Undergraduates: A Randomized Controlled Trial," *Journal of Education and Health Promotion* 9 (2020): 306.
11. N. S. King and R. M. Pringle, "Black Girls Speak STEM: Counterstories of Informal and Formal Learning Experiences," *Journal of Research in Science Teaching* 56, no. 5 (2019): 539–69.
12. King and Pringle, "Black Girls Speak STEM," 542.
13. J. M. B. Fugate, S. L. Macrine, and C. Cipriano, "The Role of Embodied Cognition for Transforming Learning," *International Journal of School and Educational Psychology* 7, no. 4 (2019): 274–88.
14. A. M. Paul, *The Extended Mind: The Power of Thinking Outside the Brain* (Eamon Dolan Books, 2021); S. A. Sloman and P. Fernbach, *The Knowledge Illusion: Why We Never Think Alone* (Penguin, 2018).
15. T. J. Smoker, C. E. Murphy, and A. K. Rockwell, "Comparing Memory for Handwriting versus Typing," *Proceedings of the Human Factors and Ergonomics Society Annual Meeting*, 53, no. 22 (2009), 1744–47.
16. For example, one study found that when first and second graders "acted out" a story they were reading by manipulating toys, they had 33 percent more recall compared to children who simply read the story aloud. A. M. Glenberg, A. B. Goldberg, and X. Zhu, "Improving Early Reading Comprehension Using Embodied CAI," *Instructional Science* 39, no. 1 (2011): 27–39.
17. R. P. Mady, T. B. Phillips, D. N. Bonter, C. Quimby, J. Borland, C. Eldermire, B. T. Walters, S. A. Parry, and M. Chu, "Engagement in the Data Collection Phase of the Scientific Process Is Key for Enhancing Learning Gains," *Citizen Science: Theory and Practice* 8, no. 1 (2023): 14.
18. C. E. Burns, "New York City Wildlife Earthwatch Research Project," Earthwatch Research Project, 2008 annual report, accessed March 28, 2024. https://www.blackrockforest.org/wp-content/uploads/2021/03/res_pub_burns_earthwatch_report_2008_forparksstaff.pdf
19. Gotham Coyote Project, "Home," accessed April 16, 2024. https://www.gothamcoyote.org/
20. Earthwatch, "Frequently Asked Questions," accessed March 28, 2024. https://earthwatch.org/about/FAQs
21. A. H. Toomey and M. C. Domroese, "Can Citizen Science Lead to Positive Conservation Attitudes and Behaviors?," *Human Ecology Review* (2013): 50–62.
22. Toomey and Domroese, "Can Citizen Science," 59.
23. A. Kollmuss and J. Agyeman, "Mind the Gap: Why Do People Act Environmentally and What Are the Barriers to Pro-Environmental Behavior?," *Environmental Education Research* 8, no. 3 (2002): 239–60.

24. See, for example, cognitive dissonance theory; L. Festinger, *A Theory of Cognitive Dissonance* (Stanford University Press, 1957).
25. D. J. Bem, "Self-Perception Theory," in *Advances in Experimental Social Psychology*, vol. 6, ed. L. Berkowitz (Elsevier, 1972), 1–62.
26. E. Aronson, C. Fried, and J. Stone, "Overcoming Denial and Increasing the Intention to Use Condoms through the Induction of Hypocrisy," *American Journal of Public Health* 81, no. 12 (1991): 1636–38.
27. Toomey and Domroese, "Can Citizen Science."
28. M. K. Halpern and K. C. Elliott, "Science as Experience: A Deweyan Model of Science Communication," *Perspectives on Science* 30, no. 4 (2022): 621–56.
29. The term *citizen science* is increasingly contentious. For a good overview of this discussion, see C. B. Cooper, C. L. Hawn, L. R. Larson, J. K. Parrish, G. Bowser, D. Cavalier, R. R. Dunn, et al., "Inclusion in Citizen Science: The Conundrum of Rebranding," *Science* 372, no. 6549 (2021): 1386–88.
30. Y. N. Golumbic, A. Motion, A. Chau, L. Choi, D. D'Silva, J. Ho, M. Nielsen, et al., "Self-Reflection Promotes Learning in Citizen Science and Serves as an Effective Assessment Tool," *Computers and Education Open* 3 (2022): 100104.
31. Zooniverse, "Projects," accessed March 28, 2024. https://www.zooniverse.org/projects
32. J. Kazmierczak, "NASA-Funded Citizen Science Project Discovers New Brown Dwarf," NASA, July 17, 2017. https://www.nasa.gov/centers-and-facilities/goddard/nasa-funded-citizen-science-project-discovers-new-brown-dwarf/
33. S. Lebeer, S. Ahannach, T. Gehrmann, S. Wittouck, T. Eilers, E. Oerlemans, S. Condori, et al., "A Citizen-Science-Enabled Catalogue of the Vaginal Microbiome and Associated Factors," *Nature Microbiology* 8, no. 11 (2023): 2183–95.
34. Julia Byrd, interview, December 19, 2023.
35. R. Bonney, J. Byrd, J. T. Carmichael, L. Cunningham, L. Oremland, J. Shirk, and A. Von Harten, "Sea Change: Using Citizen Science to Inform Fisheries Management," *Bioscience* 71, no. 5 (2021): 519–30.
36. A. H. Toomey, L. Strehlau-Howay, B. Manzolillo, and C. Thomas, "The Place-Making Potential of Citizen Science: Creating Social-Ecological Connections in an Urbanized World," *Landscape and Urban Planning* 200 (2020): 103824.
37. Participation is a buzzword that has lost much of its meaning due to its appropriation by those who would use it as a means rather than an end in itself. Some have written about the "tyranny" of participation and suggest that it is helpful to better understand motivations through which people come to

participate, as well as the different levels at which they engage. See B. Cooke and U. Kothari, *Participation: The New Tyranny?* (Zed Books, 2001).
38. J. Zeng and J. Tang, "Interaction Matters? The Impact of Volunteer-Scientist Interaction on Scientific Output of Citizen Science Projects," *Pacific Asia Conference on Information Science Proceedings* 226 (2022). https://aisel.aisnet.org/pacis2022/226
39. A. Byrne, J. Canavan, and M. Millar, "Participatory Research and the Voice-Centred Relational Method of Data Analysis: Is It Worth It?," *International Journal of Social Research Methodology* 12, no. 1 (2009): 67–77.
40. Byrne, Canavan, and Millar, "Participatory Research," 75.
41. R. M. Pateman, A. Dyke, and S. E. West, "The Diversity of Participants in Environmental Citizen Science," *Citizen Science: Theory and Practice* 6, no. 1 (2021): 9.
42. See, for example, P. M. Shepard, "Advancing Environmental Justice through Community-Based Participatory Research," *Environmental Health Perspectives* 110, suppl. 2 (2002): 139.
43. C. Garzón, B. Beveridge, M. Gordon, C. Martin, E. Matalon, and E. Moore, "Power, Privilege, and the Process of Community-Based Participatory Research: Critical Reflections on Forging an Empowered Partnership for Environmental Justice in West Oakland, California," *Environmental Justice* 6, no. 2 (2013): 71–78.
44. West Oakland Environmental Indicators Project, "Research," accessed March 28, 2024. https://woeip.org/featured-work/research/
45. Jiwan Palta, Zoom interview, January 19, 2024.
46. M. Ozgen, K. M. Farag, S. Ozgen, and J. P. Palta, "Lysophosphatidylethanolamine Accelerates Color Development and Promotes Shelf Life of Cranberries," *HortScience* 40 (2005): 127–30. Later studies established that lipids are bioactive molecules that can retard fruit degradation and promote ripening at the same time. See K. M. Farag and J. P. Palta, "Use of Lysophosphatidylethanolamine, a Natural Lipid, to Retard Tomato Leaf and Fruit Senescence," *Physiolgia Plantarum* 87 (1993): 515–24; S. B. Ryu, B. H. Karlsson, M. Ozgen, and J. P. Palta, "Inhibition of Phospholipase D by Lysophosphatidylethanolamine, a Lipid-Derived Senescence Retardant," *Proceedings of the National Academy of Science* 94 (1997): 12717–21; and Z. F. Ahmed and J. P. Palta, "Postharvest Dip Treatment with a Natural Lipid Plus Soy Lecithin Extended the Shelf Life of Banana Fruit," *Postharvest Biology and Technology* 113 (2016): 58–65.
47. M. Palta, M. V. Du Bray, R. Stotts, A. Wolf, and A. Wutich, "Ecosystem Services and Disservices for a Vulnerable Population: Findings from Urban

Waterways and Wetlands in an American Desert City," *Human Ecology* 44, no. 4 (2016): 463–78.
48. Donal Pérez Gutiérrez, WhatsApp conversation, November 2023.
49. Donal Pérez Gutiérrez, personal communication, December 15, 2023.

Chapter 7. Rethinking the "Peer" in Peer Review

1. D. J. Spiegelhalter, S. Evans, P. Aylin, and G. Murray, "Overview of Statistical Evidence Presented to the Bristol Royal Infirmary Inquiry Concerning the Nature and Outcomes of Paediatric Cardiac Surgical Services at Bristol Relative to Other Specialist Centres from 1984 to 1995," Bristol Royal Infirmary Inquiry, September 2000.
2. "Congenital Heart Disease in Children and Adults (Congenital Audit 2023)," National Institute for Cardiovascular Outcomes Research (NICOR), accessed March 28, 2024. https://www.nicor.org.uk/congenital-heart-disease-in-children-and-adults-congenital-audit/
3. "National Congenital Heart Disease Audit (Nchda)," British Congenital Cardiac Association, updated November 21, 2016. https://www.bcca-uk.org/pages/page_box_contents.asp?pageid=932&navcatid=244
4. C. Pagel, L. Rogers, K. Brown, G. Ambler, D. Anderson, D. Barron, E. Blackshaw, et al., "Improving Risk Adjustment in the PRAiS (Partial Risk Adjustment in Surgery) Model for Mortality after Paediatric Cardiac Surgery and Improving Public Understanding of Its Use in Monitoring Outcomes," *Health Services and Delivery Research* 5, no. 23 (2017): 1–164.
5. "Leeds Children's Heart Surgery Halted by 'Incomplete' Data," *BBC*, October 28, 2014. https://www.bbc.com/news/uk-england-leeds-29802038
6. Christina Pagel, Zoom interview, January 5, 2024.
7. Tracey Brown, Zoom interview, November 15, 2023.
8. Brown, interview.
9. For example, see Editors, "Communicating Risk about Children's Heart Surgery Well," *The Lancet* 387, no. 10038 (2016): 2576.
10. "Guest Post: I Hope We've Helped Others Preparing for Their Child's Heart Surgery," *Mumsnet*, February 8, 2016.
11. S. Ross, "Scientist: The Story of a Word," *Annals of Science* 18, no. 2 (1962): 65–85.
12. I am aware that this topic is controversial, but the precise date of when language evolved is not essential to my point in this book (or in talking with my students about science). For an alternative timeline, see L. Boë, T. R. Sawallis, J. Fagot, P. Badin, G. Barbier, G. Captier, L. Ménard, J. Heim, and J. Schwartz, "Which Way to the Dawn of Speech?: Reanalyzing Half a Century

of Debates and Data in Light of Speech Science," *Science Advances* 5, no. 12 (2019): eaaw3916.
13. U. N. Tasci, "How an 8th-Century Islamic Library Laid the Foundations of Modern Science," TRT World, 2021. https://www.trtworld.com/magazine/how-an-8th-century-islamic-library-laid-the-foundations-of-modern-science-12766521
14. K. Bogenschneider and T. Corbett, *Evidence-Based Policymaking: Insights from Policy-Minded Researchers and Research-Minded Policymakers* (Routledge, 2011), 337.
15. M. Bauerlein, M. Gad-el-Hak, W. Grody, B. McKelvey, and S.W. Trimble, "We Must Stop the Avalanche of Low-Quality Research," *Chronicle of Higher Education*, June 13, 2010. https://www.chronicle.com/article/we-must-stop-the-avalanche-of-low-quality-research/
16. B. Aczel, B. Szaszi, and A. O. Holcombe, "A Billion-Dollar Donation: Estimating the Cost of Researchers' Time Spent on Peer Review," *Research Integrity and Peer Review* 6, no. 14 (2021).
17. In 2010, Elsevier, one of the largest scientific publishing companies, reported a profit margin of 36 percent, higher than that of Google or Apple for the same year. Who is footing the bill for such publishing? Largely speaking, it is the tax-paying public, as well as students (and their parents), paying tuition bills, as university libraries are forced to spend millions of dollars a year to subscribe to research journals. S. Buranyi, "Is the Staggeringly Profitable Business of Scientific Publishing Bad for Science?," *The Guardian*, June 27, 2017. https://www.theguardian.com/science/2017/jun/27/profitable-business-scientific-publishing-bad-for-science
18. A. Mastroianni, "The Rise and Fall of Peer Review," *Experimental History*, December 13, 2022. https://www.experimental-history.com/p/the-rise-and-fall-of-peer-review
19. J. Grove, "Restrict Researchers to One Paper a Year, Says UCL Professor," *Times Higher Education*, November 28, 2019. https://www.timeshighereducation.com/news/restrict-researchers-one-paper-ayear-says-ucl-professor
20. For example, PeerCommunityIn (PCI) is a nonprofit organization that offers researchers peer review and publishing open access free of cost. PCI authors upload preprints of their articles, which are then evaluated by peer reviewers registered on the system. Papers are then recommended and can be published—alongside reviews—on the corresponding PCI website to be cited or recommended to PCI-friendly journals if preferred by the author. https://peercommunityin.org
21. As expressed in the words of the late Nobel Award–winning economist Eleanor

Ostrom, "The power of a theory is exactly proportional to the diversity of situations it can explain." E. Ostrom, *Governing the Commons: The Evolution of Institutions for Collective Action* (Cambridge University Press, 1990), 24.

22. R. Skibba, "The Polling Crisis: How to Tell What People Really Think," *Nature* 538, no. 7625 (2016), 304–6.
23. M. Moffett, "The "Spiral of Silence": How Pollsters Got the Colombia-FARC Peace Deal Vote So Wrong," *Vox*, October 6, 2016. https://www.vox.com/world/2016/10/6/13175608/polls-colombia-farc-peace-deal-vote-wrong
24. A. P. Christie, T. Amano, P. A. Martin, S. O. Petrovan, G. E. Shackelford, B. I. Simmons, R. K. Smith, et al., "Poor Availability of Context-Specific Evidence Hampers Decision-Making in Conservation," *Biological Conservation* 248 (2020): 108666.
25. National Commission on Excellence in Education, "A Nation at Risk: The Imperative for Educational Reform: A Report to the Nation and Secretary of Education," US Department of Education, December 1983.
26. The White House, "President Signs Landmark No Child Left Behind Education Bill," news release, January 8, 2002. https://georgewbush-whitehouse.archives.gov/news/releases/2002/01/20020108-1.html#:~:text=Not%20feel%2Dgood%20methods%2C%20not,t%20make%20a%20good%20school
27. In the language of the act itself, the phrase "research-based" is mentioned repeatedly.
28. Note that "achievement gap" language has largely been replaced with other terms, such as "opportunity gaps." See S. Y. Shukla, E. J. Theobald, J. K. Abraham, and R. M Price, "Reframing Educational Outcomes: Moving Beyond Achievement Gaps," *CBE—Life Sciences Education* 21, no. 2 (2022): es2.
29. C. Turner, "No Child Left Behind: What Worked, What Didn't," *NPR*, 2015. https://www.npr.org/sections/ed/2015/10/27/443110755/no-child-left-behind-what-worked-what-didnt
30. Vivian Tseng, Zoom interview, January 10, 2024.
31. V. Tseng and C. Coburn, "Using Evidence in the US," in *What Works Now?* (Policy Press, 2019), 351–68.
32. T. S. Dee and B. Jacob, "The Impact of No Child Left Behind on Student Achievement," *Journal of Policy Analysis and Management* 30, no. 3 (2011): 418–46.
33. J. Deke, L. Dragoset, K. Bogen, and B. Gill, "Impacts of Title I Supplemental Educational Services on Student Achievement NCEE 2012-4053," National Center for Education Evaluation and Regional Assistance, May 1, 2012.
34. C. J. Heinrich, R. H. Meyer, and G. Whitten, "Supplemental Education

Services under No Child Left Behind: Who Signs up, and What Do They Gain?," *Educational Evaluation and Policy Analysis* 32, no. 2 (2010): 273–98.
35. D. Hursh, "Exacerbating Inequality: The Failed Promise of the No Child Left Behind Act," *Race Ethnicity and Education* 10, no. 3 (2007): 295–308.
36. Tseng, interview.
37. V. Tseng, "The Next Big Leap for Research-Practice Partnerships: Building and Testing Theories to Improve Research Use," William T. Grant Foundation, December 2017.
38. P. Beier, L. J. Hansen, L. Helbrecht, and D. Behar, "A How-To Guide for Coproduction of Actionable Science," *Conservation Letters* 10, no. 3 (2017): 288–96.
39. Beier et al., "A How-To Guide," 289.
40. The films mentioned are A. Toomey, "The Park Guards of Madidi," YouTube, December 15, 2014 (https://www.youtube.com/watch?v=H-ZuJccV4as); and A. Toomey, "Indigenous Perspectives on Research," YouTube, July 6, 2015. (https://www.youtube.com/watch?v=uQd95Nq05Rk).
41. Workshops with community of San José de Uchupiamonas, November 2013, and with park guards from Pilón Lajas, October 2013. See also "What Is Research?," Science and the Community Blog, November 21, 2013. http://communitysciencebolivia.blogspot.com/2013/11/what-is-research.html
42. J. Smith-Merry, "Evidence-Based Policy, Knowledge from Experience and Validity," *Evidence and Policy* 16, no. 2 (2020): 305–16; M. S. Reed, M. Ferré, J. Martin-Ortega, R. Blanche, R. Lawford-Rolfe, M. Dallimer, and J. Holden, "Evaluating Impact from Research: A Methodological Framework," *Research Policy* 50, no. 4 (2021): 104147.
43. Pagel, interview.
44. V. Tseng, A. Bednarek, and K. Faccer, "How Can Funders Promote the Use of Research? Three Converging Views on Relational Research," *Humanities and Social Sciences Communications* 9, no. 1 (2022): 1–11.
45. A. T. Bednarek, B. Shouse, C. G. Hudson, and R. Goldburg, "Science-Policy Intermediaries from a Practitioner's Perspective: The Lenfest Ocean Program Experience," *Science and Public Policy* 43, no. 2 (2016): 291–300.
46. "Community-Engaged Scholarship for Health," accessed March 28, 2024. http://CES4Health.org
47. C. M. Jordan, S. D. Seifer, L.R. Sandmann, and S. B. Gelmon, "CES4Health. Info: Development of a Mechanism for the Peer Review and Dissemination of Innovative Products of Community-Engaged Scholarship," *International Journal of Prevention Practice and Research* 1, no. 1 (2009): 21–28.

48. *Community-Engaged Scholarship: A Faculty Handbook* (University of Louisville, Office of the Provost and Office of the Vice President for Community Engagement, n.d.), accessed March 28, 2024.
49. The Institutional Challenge Grant Program is supported by the William T. Grant Foundation, the Doris Duke Charitable Foundation, the Spencer Foundation, and the American Institutes for Research. The program encourages the reshaping of incentives in tenure and promotion to support extra-academic work by faculty. See https://wtgrantfoundation.org/funding/institutional-challenge-grant.
50. "Homepage," ideas42, accessed April 29, 2024. https://www.ideas42.org
51. Last year, I made a decision similar to Tantia's when a colleague and I decided to create our own research consulting business, called Participatory Science Solutions LLC, to support our collaborative research with environmental nonprofit and community-based organizations.
52. Piyush Tantia, personal communication, February 2024.
53. For example, colleagues and I led a collaborative research project that engaged undergraduate students and community outreach staff of an urban nonprofit organization in New York City. See A. H. Toomey, J. Smith, C. Becker, and M. Palta, "Towards a Pedagogy of Social-Ecological Collaborations: Engaging Students and Urban Nonprofits for an Ecology with Cities," *Urban Ecosystems* 26, no. 2 (2023): 425–32.

Chapter 8. The Scientist Next Door: Conversations, Communities, and Connections

1. J. C. Besley and A. H. Tanner, "What Science Communication Scholars Think about Training Scientists to Communicate," *Science Communication* 33, no. 2 (2011): 239–63.
2. A. Nerghes, B. Mulder, and J. Lee, "Dissemination or Participation? Exploring Scientists' Definitions and Science Communication Goals in the Netherlands," *PLOS ONE* 17, no. 12 (2022): e0277677.
3. F. Kearns, *Getting to the Heart of Science Communication: A Guide to Effective Engagement* (Island Press, 2021).
4. Other terms might be better fits for this type of linear activity, such as diffusion, dissemination, and even storytelling.
5. J. C. Besley and A. Dudo, *Strategic Science Communication: A Guide to Setting the Right Objectives for More Effective Public Engagement* (Johns Hopkins University Press, 2022).
6. Emily Brown (alias per request of interviewee), Zoom interview, March 19, 2024.

7. D. R. Weimer, *Summary of State Breastfeeding Laws and Related Issues* (Congressional Research Service/Library of Congress, 2005).
8. Brown, interview.
9. Brown, interview.
10. Brown, interview.
11. D. McRaney, *How Minds Change: The Surprising Science of Belief, Opinion, and Persuasion* (Oneworld Publications, 2022).
12. D. C. Beardslee and D. D. O'dowd, "The College-Student Image of the Scientist: Scientists Are Seen as Intelligent and Hard-Working but Also as Uncultured and Not Interested in People," *Science* 133, no. 3457 (1961): 997–1001.
13. Beardslee and O'dowd, "The College-Student Image," 997.
14. K. D. Finson, "Drawing a Scientist: What We Do and Do Not Know after Fifty Years of Drawings," *School Science and Mathematics* 102, no. 7 (2002): 335–45.
15. S. T. Fiske and C. Dupree, "Gaining Trust as Well as Respect in Communicating to Motivated Audiences about Science Topics," *Proceedings of the National Academy of Sciences* 111, suppl. 4 (2014): 13593–97.
16. For an example of a classic "Eureka" scene, see *Interstellar*, directed by Christopher Nolan, released November 5, 2014, by Paramount Pictures.
17. "Survey: Most Americans Cannot Name a Living Scientist or a Research Institution," *Research!America*, May 11, 2021. https://www.researchamerica.org/blog/survey-most-americans-cannot-name-a-living-scientist-or-a-research-in stitution/
18. J. Coyne, "Americans Want Science Done, but Can't Name Any Scientists or Places Where Science Is Done," *Why Evolution Is True*, January 11, 2018. https://whyevolutionistrue.com/2018/01/11/americans-want-science-done -but-cant-name-any-scientists-or-places-where-science-is-done/
19. J. L. Graves Jr., M. Kearney, G. Barabino, and S. Malcom, "Inequality in Science and the Case for a New Agenda," *Proceedings of the National Academy of Sciences* 119, no. 10 (2022): e2117831119.
20. This is particularly striking if considering that in some countries, such as the United States, rural students have higher high school graduation rates than their urban counterparts. See K. A. Schafft and A. Jackson, (Eds.), *Rural Education for the Twenty-First Century: Identity, Place, and Community in a Globalizing World* (Penn State University Press, 2011). See also L. O'Neal and A. Perkins, "Rural Exclusion from Science and Academia," *Trends in Microbiology* 29, no. 11(2021): 953–56; and K. Avendaño, "Interés Por Estudios Universitarios En Ciencia, Tecnología, Ingeniería Y Matemáticas (STEM) En Bachilleres De Tabasco," PhD diss., Universidad Juárez Autónoma De Tabasco, 2018.

21. L. M. Avery, "Rural Science Education: Valuing Local Knowledge," *Theory into Practice* 52, no. 1 (2013): 29.
22. E. Dawson, "'Not Designed for Us': How Science Museums and Science Centers Socially Exclude Low-Income, Minority Ethnic Groups," *Science Education* 98, no. 6 (2014): 981–1008.
23. E. Dawson, "Reimagining Publics and (Non) Participation: Exploring Exclusion from Science Communication through the Experiences of Low-Income, Minority Ethnic Groups," *Public Understanding of Science* 27, no. 7 (2018): 773.
24. B. Većkalov, N. Zarzeczna, J. McPhetres, F. van Harreveld, and B. T. Rutjens, "Psychological Distance to Science as a Predictor of Science Skepticism across Domains," *Personality and Social Psychology Bulletin* 50, no. 1 (2024): 18–37.
25. Većkalov et al., "Psychological Distance to Science," 34.
26. Personal communication, Natalia Zarzeczna, April 17, 2024. See also N. Zarzeczna, B. Većkalov, M. Hoffstadt, and B. Rutjens, "Psychological Distance to Science: Decreasing Distance Reduces Science Scepticism," *PsyArXiv* (preprint) (2022). doi:10.31234/osf.io/b8uej
27. A. Moreno and C. S. Olsen, "Take Rural Road Trips to Promote Science," *Nature* 543, no. 7647 (2017): 623.
28. H. Wooten, "The Roots of Cooperative Extension," *Blogs.IFAS* (blog), July 29, 2020. https://blogs.ifas.ufl.edu/orangeco/2020/07/29/the-roots-of-cooperative-extension/
29. G. R. Kremer, *George Washington Carver: In His Own Words* (University of Missouri Press, 2017), 102. This approach is similar to those promoted by Paulo Freire and other "liberation pedagogists," who would travel to places where there was little access to formal education to engage people in learning practices that built on their lived experiences. See P. Freire, *Pedagogy of the Oppressed*, 3rd ed. (Continuum, 1973).
30. BioBus, "About," accessed April 1, 2024. https://www.biobus.org/about/.
31. I first heard about Joey Rodman on David McRaney's *You Are Not So Smart* podcast (episode 222, December 12, 2021). Rodman was also interviewed on *Wheel of Randy: A Randy Newman Podcast*, hosted by Dan Wade (June 23, 2020).
32. S. Besson, "The 'Human Right to Science' qua Right to Participate in Science: The Participatory Good of Science and Its Human Rights Dimensions," *International Journal of Human Rights* (2023): 1–32.
33. C. J. Chutia, "Origin and Evolution of Ojapali: A Folk Performing Art of Assam," *Resmilitaris* 12, no. 6 (2022): 2541–51.
34. N. Borah, "Climate Change Communication through Ojapali 2: 'A Bridge

over the River,'" YouTube video, July 27, 2019. https://www.youtube.com/watch?v=y3dN4ynXpeI
35. L. McIntyre, *How to Talk to a Science Denier: Conversations with Flat Earthers, Climate Deniers, and Others Who Defy Reason* (MIT Press, 2021).
36. S. Denning, "Finding Common Ground with Climate Change Contrarians," University Corporation for Atmospheric Research, July 14, 2011. https://news.ucar.edu/4940/finding-common-ground-climate-change-contrarians
37. P. Schmid and C. Betsch, "Effective Strategies for Rebutting Science Denialism in Public Discussions," *Nature Human Behaviour* 3, no. 9 (2019): 931–39.
38. McIntyre, *How to Talk to a Science Denier*, 8.
39. J. Medina, "A New Strategy to Persuade Voters: Listen Carefully. And Don't Hurry," *New York Times*, October 20, 2021. https://www.nytimes.com/2021/10/20/us/politics/minneapolis-deep-canvassing.html
40. J. L. Kalla and D. E. Broockman, "Reducing Exclusionary Attitudes through Interpersonal Conversation: Evidence from Three Field Experiments," *American Political Science Review* 114, no. 2 (2020): 410–25.
41. Neighbours United, "Deep Engagement," accessed on April 12, 2024. https://neighboursunited.org/campaign/deep-engagement/
42. *Deep Canvassing on Climate—The Power of Listening to Persuade*, video, Yale Program on Climate Change Communication, December 1, 2023. https://www.youtube.com/watch?v=xQD1QruIr1U
43. Explore Your Trail, "2023 Children's Lead Levels Continue to Decline," news release, December 7, 2023. https://trail.ca/en/news/2023-children-s-lead-levels-continue-to-decline.aspx
44. Canvassers had 1,181 conversations, of which 796 were considered "complete" conversations (the canvassers went through the entire script); 506 residents (42.8 percent) who were canvassed signed a petition in support of climate action. The methodology is available for download at neighboursunited.org/campaign/deep-engagement/#playbook-toolkit.
45. *Deep Canvassing on Climate*. See from time stamp 38:00.
46. Besley and Dudo, *Strategic Science Communication*.
47. See J. C. Besley and M. Nisbet, "How Scientists View the Public, the Media and the Political Process," *Public Understanding of Science* 22, no. 6 (2013): 644–59.
48. John Besley, personal communication, January 2024.
49. *Deep Canvassing on Climate*.
50. S. Kleinberg and A. H. Toomey, "The Use of Qualitative Research to Better Understand Public Opinions on Climate Change," *Journal of Environmental Studies and Sciences* 13 (2023): 367–75.

51. "Julia Minson's Story," *Hidden Brain*, podcast, aired July 11, 2023. https://hiddenbrain.org/unsunghero/julia-minsons-story/
52. "Julia Minson's Story." Quote begins at time stamp 2:30.
53. Science communication scholars John Besley and Anthony Dudo point out that when people are given an opportunity to have their perspective heard, they often feel more accepting of a resulting decision (even if they disagree with the decision). Besley and Dudo, *Strategic Science Communication*.
54. F. Kearns, *Getting to the Heart of Science Communication: A Guide to Effective Engagement*. (Island Press, 2021): 85.

Chapter 9. The Skeptic in the Mirror: The Essential Role of Uncertainty in Science

1. M. Jackson, *Uncertain: The Wisdom and Wonder of Being Unsure* (Prometheus, 2023).
2. F. Schönweitz, J. Eichinger, J. M. Kuiper, F. Ongolly, W. Spahl, B. Prainsack, and B. M. Zimmermann, "The Social Meanings of Artifacts: Face Masks in the Covid-19 Pandemic," *Front Public Health* 10 (2022): 829904.
3. L. H. Kahane, "Politicizing the Mask: Political, Economic and Demographic Factors Affecting Mask Wearing Behavior in the USA," *Eastern Economic Journal* 47, no. 2 (2021): 163–83.
4. C. Wu, "Brisbane Man 'Suffers Heart Attack' after Being Arrested for Not Wearing a Mask," *SkyNews*, August 4, 2021. https://www.skynews.com.au/australia-news/coronavirus/brisbane-man-suffers-heart-attack-after-being-arrested-for-not-wearing-a-mask-while-exercising-in-lockdown/news-story/cc679efd8d27d7cbf5792eef75b6e857
5. S. Ankel, "An Elementary School Teacher Was Beaten on the First Day of the New Semester by a Father Angry about a Mask Mandate," *Business Insider*, August 14, 2021. https://www.businessinsider.com/california-parent-beats-teacher-over-mask-dispute-first-school-day-2021-8
6. "Germans Shocked by Killing of Cashier after COVID-19 Mask Row," *Reuters*, September 21, 2021. https://www.reuters.com/world/europe/germans-shocked-by-killing-of-cashier-after-COVID-19-mask-row-2021-09-21/
7. J. Wong and E. Claypool, "Narratives, Masks and COVID-19: A Qualitative Reflection," *Qualitative Social Work* 20, no. 1–2 (2021): 206–13.
8. T. Jefferson, L. Dooley, E. Ferroni, L. A. Al-Ansary, M. L. van Driel, G. A. Bawazeer, M. A. Jones, et al., "Physical Interventions to Interrupt or Reduce the Spread of Respiratory Viruses," *Cochrane Database of Systematic Reviews*, no. 1 (2023).
9. The interview with Tom Jefferson was conducted by reporter Maryanne

Demasi in February 2023. As described in B. Stephens, "Opinion: The Mask Mandates Did Nothing. Will Any Lessons Be Learned?," *New York Times*, February 21, 2023. https://www.nytimes.com/2023/02/21/opinion/do-mask-mandates-work.html

10. For examples, see N. Swaminathan, "The Fear Factor: When the Brain Decides It's Time to Scram," *Scientific American*, August 24, 2007; K. Padavic-Callaghan, "Identical Quantum Particles Pass Practicality Test," *Scientific American*, September 28, 2020. https://www.scientificamerican.com/article/identical-quantum-particles-pass-practicality-test/; and R. Kaufman, "Youngest Planet Confirmed; Photos Show It Grew up Fast," *National Geographic*, June 12, 2010. https://www.nationalgeographic.com/science/article/100610-youngest-planet-exoplanet-space-science

11. For examples, see A. M. Potapov, J. Drescher, K. Darras, A. Wenzel, N. Janotta, R. Nazarreta, et al., "Rainforest Transformation Reallocates Energy from Green to Brown Food Webs," *Nature* 627, no. 8002 (2024): 116–22.; X. Jiang, Z. Qijing Zheng, Z. Lan, W. A. Saidi, X. Ren, and J. Zhao, "Real-Time *GW*-BSE Investigations on Spin-Valley Exciton Dynamics in Monolayer Transition Metal Dichalcogenide," *Science Advances* 7, no. 10 (2021): eabf3759; H. Huang, R. Kueng, G. Torlai, V. V. Albert, and J. Preskill, "Provably Efficient Machine Learning for Quantum Many-Body Problems," *Science* 377, no. 6613 (2022): eabk3333; and S. Lomoio, R. Willen, W. Kim, K. Z. Ho, E. K. Robinson, D. Prokopenko, M. E. Kennedy, R. E. Tanzi, and G. Tesco, "*Gga3* Deletion and a *GGA3* Rare Variant Associated with Late Onset Alzheimer's Disease Trigger BACE1 Accumulation in Axonal Swellings," *Science Translational Medicine* 12, no. 570 (2020): eaba1871.

12. B. M. Gurbaxani, A. N. Hill, and P. Patel, "Unpacking Cochrane's Update on Masks and COVID-19," *American Journal of Public Health* 113, no. 10 (2023): 1074–78.

13. See original studies: J. Abaluck, L. H. Kwong, A. Styczynski, A. Haque, M. A. Kabir, E. Bates-Jefferys, E. Crawford, et al., "Impact of Community Masking on Covid-19: A Cluster-Randomized Trial in Bangladesh," *Science* 375, no. 6577 (2022): eabi9069; H. Bundgaard, J. S. Bundgaard, D. E. T. Raaschou-Pedersen, C. von Buchwald, T. Todsen, J. B. Norsk, M. M. Pries-Heje, et al., "Effectiveness of Adding a Mask Recommendation to Other Public Health Measures to Prevent Sars-Cov-2 Infection in Danish Mask Wearers: A Randomized Controlled Trial," *Annals of Internal Medicine* 174, no. 3 (2021): 335–43.

14. Observational studies were included in the first four versions of the review, but not in the last two versions (published in 2020 and 2023). See version history

at T. Jefferson, L. Dooley, E. Ferroni, L. A. Al-Ansary, M. L. van Driel, G. A. Bawazeer, M. A. Jones, et al., "Physical Interventions to Interrupt or Reduce the Spread of Respiratory Viruses," *Cochrane Database of Systematic Reviews*, no. 1 (2023).
15. N. Oreskes, "What Went Wrong with a Highly Publicized COVID-19 Mask Analysis?," *Scientific American*, November 1, 2023. https://www.scientificamerican.com/article/what-went-wrong-with-a-highly-publicized-COVID-19-mask-analysis/
16. T. Jefferson, L. Dooley, E. Ferroni, L. A. Al-Ansary, M. L. van Driel, G. A. Bawazeer, M. A. Jones, et al., "Physical Interventions to Interrupt or Reduce the Spread of Respiratory Viruses," *Cochrane Database of Systematic Reviews*, no. 1 (2023).
17. "Doctor over Mask Study: People are 'Drunk with Certainty' around Mask," CNN, September 9, 2023. https://www.cnn.com/videos/health/2023/09/09/smr-mask-study-doctor-uncertainty.cnn
18. P. D. Ziakas and E. Mylonakis, "Public Interest Trends for Covid-19 and Alignment with the Disease Trajectory: A Time-Series Analysis of National-Level Data," *PLOS Digit Health* 2, no. 6 (2023): e0000271.
19. J. Symanzik, "The Good, the Bad, and the Ugly Coronavirus Graphs," Southwest Michigan Chapter of the American Statistical Association (ASA), virtual, January 7, 2021.
20. M. D. Baird, D. G. Groves, O. A. Osoba, A. M. Parker, R. Sanchez, and C. M. Setodji, "Don't Make the Pandemic Worse with Poor Data Analysis," commentary, RAND, May 6, 2020. https://www.rand.org/blog/2020/05/dont-make-the-pandemic-worse-with-poor-data-analysis.html
21. M. L. Urrutia, "Covid Data Is Complex and Changeable——Expecting the Public to Heed It as Restrictions Ease Is Optimistic," *The Conversation*, July 16, 2021. https://theconversation.com/covid-data-is-complex-and-changeable-expecting-the-public-to-heed-it-as-restrictions-ease-is-optimistic-164609
22. C. Lee, T. Yang, G. D. Inchoco, G. M. Jones, and A. Satyanarayan, "Viral Visualizations: How Coronavirus Skeptics Use Orthodox Data Practices to Promote Unorthodox Science Online," Proceedings of the 2021 CHI Conference on Human Factors in Computing Systems, 607 (2021): 1–18.
23. Lee et al., 2.
24. D. M. Kahan, E. Peters, E. C. Dawson, and P. Slovic, "Motivated Numeracy and Enlightened Self-Government," *Behavioural Public Policy* 1 (2017): 54–86.
25. For example, see D. M. Kahan, E. Peters, M. Wittlin, P. Slovic, L. Larrimore Ouellette, D. Braman, and G. Mandel, "The Polarizing Impact of Science

Literacy and Numeracy on Perceived Climate Change Risks," *Nature Climate Change* 2, no. 10 (2012): 732–35.
26. L. McIntyre, *How to Talk to a Science Denier: Conversations with Flat Earthers, Climate Deniers, and Others Who Defy Reason* (MIT Press, 2021).
27. E. C. Anderson, R. N. Carleton, M. Diefenbach, and P. K. J. Han, "The Relationship between Uncertainty and Affect," *Frontiers in Psychology* 10 (2019): 2504.
28. S. Chaiken and S. Yates, "Affective-Cognitive Consistency and Thought-Induced Attitude Polarization," *Journal of Personality and Social Psychology* 49, no. 6 (1985): 1470.
29. N. Light, P. M. Fernbach, N. Rabb, M.V. Geana, and S. A. Sloman, "Knowledge Overconfidence is Associated with Anti-Consensus Views on Controversial Scientific Issues," *Science Advances* 8, no. 7 (2022): eabo0038.
30. Often referred to as the Dunning-Kruger effect based on the paper by J. Kruger and D. Dunning, "Unskilled and Unaware of It: How Difficulties in Recognizing One's Own Incompetence Lead to Inflated Self-Assessments," *Journal of Personality and Social Psychology* 77, no. 6 (1999): 1121.
31. P. M. Fernbach, N. Light, S. E. Scott, Y. Inbar, and P. Rozin, "Extreme Opponents of Genetically Modified Foods Know The Least but Think They Know the Most," *Nature Human Behaviour* 3, no. 3 (2019): 251–56.
32. N. Light, P. M. Fernbach, N. Rabb, M. V. Geana, and S. A. Sloman, "Knowledge Overconfidence Is Associated with Anti-Consensus Views on Controversial Scientific Issues," *Science Advances* 8, no. 29 (2022): eabo0038.
33. S. A. Sloman and P. Fernbach, *The Knowledge Illusion: Why We Never Think Alone* (Penguin, 2018).
34. It is important that this be done with kindness and a sense of shared discovery rather than a "gotcha" kind of approach; otherwise, it will just put people on the defensive. See McIntyre, *How to Talk to a Science Denier*.
35. D. M. Kahan, A. Landrum, K. Carpenter, L. Helft, and K. Hall Jamieson, "Science Curiosity and Political Information Processing," *Political Psychology* 38 (2017): 179–99.
36. David Spiegelhalter, Zoom interview, January 9, 2024.
37. D. Michaels, *Doubt Is Their Product: How Industry's Assault on Science Threatens Your Health* (Oxford University Press, 2008), xi.
38. Michaels, *Doubt Is Their Product*, x.
39. M. De Witte, "How Does Uncertainty in Scientific Predictions Affect Credibility?," *On Our Planet* (newsletter), Stanford Doerr School of Sustainability, October 14, 2019. https://sustainability.stanford.edu/news/how-does-uncertainty-scientific-predictions-affect-credibility

40. A. M. Van Der Bles, S. Van Der Linden, A. L. J Freeman, J. Mitchell, A. B. Galvao, L. Zaval, and D. J. Spiegelhalter, "Communicating Uncertainty about Facts, Numbers and Science," *Royal Society Open Science* 6, no. 5 (2019): 181870.
41. A. Gustafson and R. E. Rice, "A Review of the Effects of Uncertainty in Public Science Communication," *Public Understanding of Science* 29, no. 6 (2020): 614–33.
42. C. Sagan, *The Demon-Haunted World: Science as a Candle in the Dark* (Ballantine, 1996), 22.
43. S. Firestein, *Ignorance: How It Drives Science* (Oxford University Press, 2012), 4.
44. U. Alon, "Ted Talk: Why Science Demands a Leap into the Unknown," 2013. https://www.ted.com/talks/uri_alon_why_science_demands_a_leap_into_the_unknown?language=en
45. Alon, "Ted Talk." Quote begins at time stamp 11:16.
46. H.-Y. Hong and X. Lin-Siegler, "How Learning about Scientists' Struggles Influences Students' Interest and Learning in Physics," *Journal of Educational Psychology* 104, no. 2 (2012): 469.
47. K. H. Jamieson, M. McNutt, V. Kiermer, and R. Sever. "Signaling the Trustworthiness of Science," *Proceedings of the National Academy of Sciences* 116, no. 39 (2019): 19231–36.
48. N. Oreskes, *Why Trust Science?* (Princeton University Press, 2019).
49. N. Oreskes, "Scientists Should Admit They Bring Personal Values to Their Work," *Scientific American*, April 1, 2021. https://www.scientificamerican.com/article/scientists-should-admit-they-bring-personal-values-to-their-work/
50. K. C. Elliott, *A Tapestry of Values: An Introduction to Values in Science* (Oxford University Press, 2017).
51. Gurbaxani, Hill, and Patel, "Unpacking Cochrane's Update."
52. B. Shine, K. L. Brown, C. Felts, and T. Mitchell, "Student and Faculty Perceptions of the Impact of Masks on Student Learning and Communication in the Classroom," *Midwest Social Sciences Journal* 26, no. 1 (2023): 8.
53. A personal note about my own values and biases here: although personally vaccinated against COVID-19, I have several close friends who are unvaccinated and unwilling to get the vaccine. During the pandemic, I had many long and intense arguments with some of them, even engaging in some of the shaming behavior I critiqued in chapter 2 and now regret. The process of writing this book and reflecting on these arguments has given me a different perspective on the issue. On one hand, I am more aware of my own lack of expertise in terms

of the science of vaccines, masking, and other health measures taken during the pandemic and thus am more inclined than ever to trust the experts. On the other, I am increasingly aware that many decisions made during the pandemic went beyond the science and were largely about prioritizing some values over others.

54. A. Stevens, "Governments Cannot Just 'Follow the Science' on COVID-19," *Nature Human Behaviour* 4, no. 6 (2020): 560–60.
55. See D. Leonhardt, "Follow the Science?," *New York Times*, February 11, 2022. https://www.nytimes.com/2022/02/11/briefing/COVID-19-cdc-follow-the-science.html

Chapter 10. In the Belly of the Beast: Scientists, Policy-Making, and Advocacy

1. See C. Rudder, "Let's March to Stress the Value of Science for the Public Good, Not to Engage in Partisan Politics," *Proceedings of the National Academy of Sciences* 114, no. 15 (2017): 3784–86.
2. G. E. Garrard, F. Fidler, B. C. Wintle, Y. E. Chee, and S. A. Bekessy, "Beyond Advocacy: Making Space for Conservation Scientists in Public Debate," *Conservation Letters* 9, no. 3 (2016): 208–12.
3. C. Sagan, "Speaking Out," *Science* 260, no. 5116 (1993): 1861–61.
4. There have always been scientists who have bucked the trend, however. For example, John Snow, considered to be a pioneer of modern epidemiology due to his role in understanding waterborne diseases, spent the latter years of his life trying, often unsuccessfully, to convince public officials to reduce the contamination of drinking water from cesspools and sewers. His one tangible success—to get officials to remove the handle of a water pump that he deemed responsible for an outbreak of cholera—was short-lived, as officials replaced the handle shortly afterward. Another prominent scientist, Rachel Carson, is well known for her pivotal role in advancing the environmental movement.
5. N. Howe, "'Stick to the Science': When Science Gets Political," *Nature*, November 3, 2020. https://www.nature.com/articles/d41586-020-03067-w
6. R. Abramoff, "I'm a Scientist Who Spoke Up about Climate Change. My Employer Fired Me," *New York Times*, January 10, 2023. https://www.nytimes.com/2023/01/10/opinion/scientist-fired-climate-change-activism.html
7. E. Pain, "How Scientists Can Influence Policy," *Science,* Feburary 14, 2014. https://www.science.org/content/article/how-scientists-can-influence-policy
8. J. C. Besley and M. Nisbet, "How Scientists View the Public, the Media and the Political Process," *Public Understanding of Science* 22, no. 6 (2013): 644–59.

9. H. Miller and A. Berezow, "March for Science Was Just an Excuse to Attack Trump and Republicans," *Fox News*, April 19, 2018. https://www.foxnews.com/opinion/march-for-science-was-just-an-excuse-to-attack-trump-and-republicans
10. A. D. Ross, R. Struminger, J. Winking, and K. R. Wedemeyer-Strombel, "Science as a Public Good: Findings from a Survey of March for Science Participants," *Science Communication* 40, no. 2 (2018): 228–45.
11. R. A. Pielke Jr., *The Honest Broker: Making Sense of Science in Policy and Politics.* (Cambridge University Press, 2007).
12. M. P. Nelson and J. A. Vucetich, "On Advocacy by Environmental Scientists: What, Whether, Why, and How," *Conservation Biology* 23, no. 5 (2009): 1090–101.
13. In 2023, the course was renamed and is now the Animal Advocacy Clinic. See https://www.pace.edu/dyson/faculty-and-research/research-centers-and-initiatives/animal-policy-project
14. A. Treves, M. Krofel, and J. McManus, "Predator Control Should Not Be a Shot in the Dark," *Frontiers in Ecology and the Environment* 14, no. 7 (2016): 380–88.
15. T. Bach, B. Niklasson, and M. Painter, "The Role of Agencies in Policy-Making," *Policy and Society* 31, no. 3 (2012): 183–93.
16. Michelle Land, interview, June 20, 2023.
17. P. Cairney and K. Oliver, "How Should Academics Engage in Policymaking to Achieve Impact?," *Political Studies Review* 18, no. 2 (2020): 228–44.
18. Michelle Land as quoted in A. Luby, "State Bans Wildlife Killing Contests; Pace Students Support Measure," *Examiner News*, January 10, 2024. https://www.theexaminernews.com/state-bans-wildlife-killing-contests-pace-students-support-measure/
19. W. J. Sutherland and C. F. Wordley, "Evidence Complacency Hampers Conservation," *Nature Ecology and Evolution* 1, no. 9 (2017): 1215–16; M. C. Evans, F. Davila, A. H. Toomey, and C. Wyborn, "Embrace Complexity to Improve Conservation Decision Making," *Nature Ecology and Evolution* 1, no. 11 (2017): 1588–88.
20. P. Cairney, A. Boaz, and K. Oliver, "Translating Evidence into Policy and Practice: What Do We Know Already, and What Would Further Research Look Like?," *BMJ Quality and Safety*, March 22, 2023.
21. P. Cairney and R. Kwiatkowski, "How to Communicate Effectively with Policymakers: Combine Insights from Psychology and Policy Studies," *Palgrave Communications* 3, no. 1 (2017): 2.

22. P. Cairney, "Three Habits of Successful Policy Entrepreneurs," *Policy and Politics* 46, no. 2 (2018): 199–215.
23. Cairney and Kwiatkowski, "How to Communicate Effectively."
24. "What Is Evidence Week?," Sense about Science, accessed April 1, 2024. https://senseaboutscience.org/what-is-evidence-week/
25. C. Whittall, "Imperial Aerodynamics Expert Briefs MPs on Future Vehicle Design," Imperial, November 28, 2022. https://www.imperial.ac.uk/news/241916/imperial-aerodynamics-expert-briefs-mps-future/
26. Tracey Brown, Zoom interview, November 15, 2023.
27. Rebecca Adler Miserendino, Zoom interview, December 20, 2023.
28. Land, interview.
29. Land, interview.
30. B. D. Gold, "The Aldo Leopold Leadership Program: Training Environmental Scientists to Be Civic Scientists," *Science Communication* 23, no. 1 (2001): 41–49.
31. Cairney, "Three Habits of Successful Policy Entrepreneurs."
32. K. Bogenschneider and T. Corbett, *Evidence-Based Policymaking: Insights from Policy-Minded Researchers and Research-Minded Policymakers* (Routledge, 2010).
33. G. Whitesides, "Learning from Success: Lessons in Science and Diplomacy from the Montreal Protocol," *Science and Diplomacy* 9, no. 2 (2020): 1–13.
34. K. Litfin, *Ozone Discourses: Science and Politics in Global Environmental Cooperation* (Columbia University Press, 1994), 198.
35. D. C. Rose, N. Mukherjee, B. I. Simmons, E. R. Tew, R. J. Robertson, A. B. Vadrot, R. Doubleday, and W.J. Sutherland, "Policy Windows for the Environment: Tips for Improving the Uptake of Scientific Knowledge," *Environmental Science and Policy* 113 (2020): 47–54.
36. M. C. Nisbet and C. Mooney, "Framing Science," *Science* 316 (2007).
37. P. Cairney, *The Politics of Evidence-Based Policy Making* (Springer, 2016). See especially chap. 3.
38. A. M. Kusmanoff, F. Fidler, A. Gordon, G. E. Garrard, and S. A. Bekessy, "Five Lessons to Guide More Effective Biodiversity Conservation Message Framing," *Conservation Biology* 34, no. 5 (2020): 1131–41.
39. Land, interview.
40. Land, interview.
41. "Zones of Agreement," *Blue Mountains Forest Partners*, accessed April 13, 2024. https://bluemountainsforestpartners.org/work/zones-of-agreement/
42. S. V. Wing, "Logging Plans for Some Eastern Oregon Forests May Now Prioritize Wildlife," Oregon Public Broadcasting, May 22, 2023. https://

www.opb.org/article/2023/05/22/logging-plans-for-some-eastern-oregon-forests-may-now-prioritize-wildlife/

43. A. Scott, "What Happens When Loggers and Environmentalists Work Together?," Oregon Public Radio, October 26, 2020. https://www.opb.org/article/2020/10/26/what-happens-when-loggers-and-environmentalists-work-together/

44. A. Scott, "A Way Forward," October 3, 2020, episode 7 in *Timber Wars*, podcast. https://www.opb.org/article/2020/10/03/environmentalist-loggers-common-ground/

45. Anuj Shah, zoom conversation, December 4, 2023.

46. Anuj Shah, personal communication, February 2024.

47. O. Dube, S. J. MacArthur, and A. K. Shah, *A Cognitive View of Policing* (National Bureau of Economic Research, 2023).

48. D. C. Rose, N. Mukherjee, B. I. Simmons, E. R. Tew, R. J. Robertson, A. B. Vadrot, R. Doubleday, and W.J. Sutherland, "Policy Windows for the Environment: Tips for Improving the Uptake of Scientific Knowledge," *Environmental Science and Policy* 113 (2020): 47–54; P. Gluckman, "Policy: The Art of Science Advice to Government," *Nature* 507, no. 7491 (2014): 163–65.

49. J. Bandola-Gill, "Between Relevance and Excellence? Research Impact Agenda and the Production of Policy Knowledge," *Science and Public Policy* 46, no. 6 (2019): 895–905. See also C. Weiss, "Research for Policy's Sake: The Enlightenment Function of Social Research," *Policy Analysis* 3 (1977): 531–45.

50. D. W. Cash and P. G. Belloy, "Salience, Credibility and Legitimacy in a Rapidly Shifting World of Knowledge and Action," *Sustainability* 12, no. 18 (2020): 7376.

51. R. A. Pielke Jr., *The Honest Broker: Making Sense of Science in Policy and Politics*. (Cambridge University Press, 2007).

52. W. G. Wells Jr., *Working with Congress: A Practical Guide for Scientists and Engineers* (American Institute of Physics, 1994).

53. P. Cairney and K. Oliver, "How Should Academics Engage in Policymaking to Achieve Impact?," *Political Studies Review* 18, no. 2 (2020): 228–44. See also Z. Zevallos, "Protecting Activist Academics against Public Harassment," *The Other Sociologist* 6, July 6, 2017. https://othersociologist.com/2017/07/06/activist-academics-public-harassment/

54. M. Torres and S. Branford, "Intimidation of Brazil's Enviro Scientists, Academics, Officials on Upswing," Mongabay, April 8, 2021. https://news.mongabay.com/2021/04/intimidation-of-brazils-enviro-scientists-academics-officials-on-upswing/

55. Extension work is a good example of an application-focused research position, but there are a very limited number of such positions in most countries in the world. China is a notable exception with more than six hundred thousand positions (compare that with fewer than three thousand positions in the United States). See B. E. Swanson, and K. Davis, "Status of Agricultural Extension and Rural Advisory Services Worldwide: Summary Report," Extension Education Worldwide, Global Forum for Rural Advisory Services (2014).
56. D. H. Guston, *Boundary Organizations in Environmental Policy and Science: An Introduction* (Sage, 2001): 399–408.
57. A. T. Bednarek, B. Shouse, C. G. Hudson, and R. Goldburg, "Science-Policy Intermediaries from a Practitioner's Perspective: The Lenfest Ocean Program Experience," *Science and Public Policy* 43, no. 2 (2016): 291–300.
58. See also B. Holmes, G. Scarrow, and M. Schellenberg, "Translating Evidence into Practice: The Role of Health Research Funders," *Implementation Science* 7, no. 1 (2012): 1–10.
59. J. H. Phoenix, L.G. Atkinson, and H. Baker, "Creating and Communicating Social Research for Policymakers in Government," *Palgrave Communications* 5, no. 1 (2019): 1–11.
60. Also in the United States, the Intergovernmental Personnel Act creates short-term opportunities for researchers to spend time in government.
61. C. Hill, "Geopolicy: Science for Policy Internships and Traineeships—a Regularly Updated List," November 24, 2023. https://blogs.egu.eu/geolog/2023/11/24/geopolicy-science-for-policy-internships-traineeships-an-incomplete-list/
62. American Institute of Biological Sciences, "Policy and Advocacy Services," accessed April 4, 2024. https://www.aibs.org/policy/
63. Some examples include the Animal Welfare Institute in Washington, DC, the Rockefeller Foundation in New York City, and the Center for Productive Rights in Los Angeles.
64. B. Iasevoli, "A Glut of Ph.D.s Means Long Odds of Getting Jobs," National Public Radio, February 27, 2015. https://www.npr.org/sections/ed/2015/02/27/388443923/a-glut-of-ph-d-s-means-long-odds-of-getting-jobs#:~:text=D.s%20Means%20Long%20Odds%20Of%20Getting%20Jobs%20%3A%20NPR%20Ed%20Only,universities%20keep%20churning%20them%20out
65. B. Alberts, "New Career Paths for Scientists," *Science* 320, no. 5874 (2008): 289.
66. Michelle Land, personal communication, February 2024.

Conclusion and Acknowledgments. From Boldly Going to Steadily Engaging

1. *Star Trek: Lower Decks*, "Second Contact," season 1, episode 1, aired August 6, 2020. Quote begins at time stamp 00:40.
2. L. Beck, "Firsting in Discovery and Exploration History," *Terrae Incognitae* 49, no. 2 (2017): 109.
3. See also M. Liboiron, "Firsting in Research," *Discard Studies*, January 18, 2021. https://discardstudies.com/2021/01/18/firsting-in-research/
4. Never mind if that new species was known for generations to the people native to the region. Science demands that the finding be made official by publishing a paper about it. There are countless examples; a recent one is M. Ives and H. Nindita, "A Plant That Flowers Underground Is New to Science, but Not to Borneo," *New York Times*, January 20, 2024.
5. As discussed in chapter 3. See also G. W. Allport, *The Nature of Prejudice* (Addison-Wesley, 1954), 263.
6. Roma Solomon, Zoom interview, January 4, 2024.
7. S. Shapin, "Science and the Public," in *Companion to the History of Modern Science*, ed. R. C. Olby (Routledge, 1990).

Index

access points, 48–49, 61–66
actionable and policy-relevant science. *See also* impact of research
   Children's Heart Surgery Outcomes website, 139–43, 154
   evidence-based policy movement, 148–51
   evidence uses vs. bases, 151–54
   funding and incentive barriers, 154–57
   peer review and extra-academic communication, 143–46
   scientific validity, external and internal, 147–48
Aiken, Gerald Taylor, 99–100
Ajwang', Robert, 98–99
Alberts, Bruce, 120
Allport, Gordon, 66–67
Alon, Uri, 196–97
Alzheimer's research, 87
Amazon, Bolivian, 51–54, 57–61
American Institute of Biological Sciences, 220
Ames, Bruce, 202
Ansari, Aziz, 41

Balmford, Andrew, 57
Beck, Lauren, 222
Beck, Ulrich, 105–6
Bem, Daryl, 127
Besley, John, 176
BioBus initiative, 171
Blair, Tony, 40
blanket consent, 108–9
Blue Mountain Forest Partners, 215–16
Bogenschneider, Karen, 144
Bolivian Amazon, 51–54, 57–61
Bomgardner, Matt, 88–89, 147
Borah, Nabanita, 171–72
boundary organizations, 218–20
Bowman, Leslie, 89–90
brain, human, 18–24, 43
Brown, Emily, 164–66
Brown, Tracey, 141–42
Burg, David, 161–62, 178

Burgess, Montana, 175
Bush, George W., 148–49
Bush, Vannevar, 84
Button, Cat, 99–100
Byrd, Julia, 129–30, 132
Byrne, Anne, 131–32

Cairney, Paul, 207–8
canvassing, deep, 174–75
Carver, George Washington, 170–71
Centola, Damon, 44–45
certainty. *See* uncertainty
Chain, Bill, 73–75, 81, 88–89
Cheatum, Molly, 73, 88
Chicago Police Department, 216
Children's Heart Surgery Outcomes website, 139–43, 154
choice
    cultural competency and equitable collaborations, 101–4
    location, 98–101
    not doing something, 114–15
    parachute science and, 95–97, 113–14
    permissions and consent, 104–9
    policy-making and, 203
citizen science, 128–32
Civic Laboratory for Environmental Action Research (CLEAR), 106–7
Clark, Jasmine, 203
climate change
    fact-based appeals, political polarization, and skepticism, 30–31
    Gore's *An Inconvenient Truth*, 25–27
    Heartland Institute conference, 172–73
    Neighbours United, 174–75
    refugees of, 16
    Tangier Island and climate skepticism, 15–17, 24–29
Cochrane report, 182–83, 185–87
Colbert, Stephen, 17
collaboration, 85–87, 101–4, 111–12

communication, meaning of, 163
Community-Engaged Scholarship for Health website, 155
community peer review (CLEAR), 106–7
compensation of local assistants, 103–4
complex contagion scenarios, 44–47
consent, 104–9
conservation science (conservation biology), 54–57
contact hypothesis, 66–67, 222
conversations with the public. *See also* listening; question-asking
    distance vs. proximity to science, 170
    intentionality, 176
    meaning of communication, 163
    mobile laboratories, 170–71
    right to science and, 171–72
    science deserts and, 169
    scientists as bad communicators, 3, 162–63
    skepticism, techniques for, 172–75
    social confirmation and, 164–67
    stereotypes of scientists, 167–68
    unplanned, 133–37
    value of, 161–62, 177–78
Copa Alvaro, María Eugenia, 62–66
Corbett, Thomas, 144
COVID-19
    masking, 179–83, 185–90, 198–99
    vaccine hesitancy, 36, 38–39, 165–66, 192–93
Cowling, Richard, 57
Criswell, Lucas, 77
Cronin, John, 204
cultural competency, 101–3

data collection
    citizen science, 128–33
    in Madidi, Bolivia, 57, 62
    New York City Wildlife Earthwatch Research Projects, 124–126

quadrat and transect sampling, 117–20, 184
science as a verb, 120–24
unplanned conversations with the public, 133–37
data interpretation, 18–25, 58–68, 131–32, 139–41, 187–90
data literacy, 187–90
Dawson, Emily, 169
decision making, 60–61, 208–9
deep canvassing, 174–75
deficit model, 56
deliberation, access to, 113
Denning, Scott, 172–73
distance to science, 170
diversity, human, 87
dogma, 86–87
Don't Look Up (film), 13–15, 31
Douglas, Maggie, 75–81, 91–93
"the dress" phenomena, 22–23
Dube, Oeindrila, 216

Earthwatch Institute, 125
Edmondson, Laura, 98–99
education
  experiential learning, 122–23, 157
  No Child Left Behind Act, 148–50
  science as a verb, 120–24
Ehrenfeld, David, 54
embodied cognition, 123
encounter, spaces of, 67–68, 136, 219
Epstein, Steven, 113
Eskridge, James "Ooker," 17, 25, 27–29
ethics
  compensation of local assistants, 103–4
  "do no harm," 95, 113
  permissions and consent, 104–9
  research as "neutral" and, 96
evidence-based policy movement, 148–51. *See also* policy-making, evidence-based

Evidence Week, 209
evolution, human, 19–20

farming
  context dependence and types of knowledge, 88–91
  cranberry and potato farmers, Wisconsin, 134–36
  extension programs, 75–76
  no-till farmers, slugs, and neonicotinoid insecticides, 73–81
  Sustainable Agricultural Research and Education (SARE), 91–92
  twenty-mile rule, 81
Fernbach, Philip, 20
fetal tissue research and consent, 108–9
filmmaking, 152–53
Firestein, Stuart J., 195–96
firsting, 222
fish release study, 130
foldscopes, 111
Freire, Paulo, 71
frugal science, 110–11
funding
  Alzheimer's research, 87
  cultural competency and, 102–3
  equitable collaboration and, 101
  extra-academic communication and, 154–57
  impact and, 68, 84–85
  incentives and, 155–56, 223
  Institutional Challenge Grants, 156
  methodology and frugal science, 110–11
  models of, 91–92
  research-practice gap and, 55–56, 75, 90–91

Giddens, Anthony, 48
Gore, Al, 25–30
Gotham Coyote Project, 125
groupthink, 86–87, 91

Hansen, James, 3
"hardly reached" communities, 35, 48
Harris, Jordan, 211
Havasupai tribe, 102
heuristics, 21
HIV/AIDS, 112–13, 127
Hochul, Kathy, 204
honesty, 193, 197–98
hooks, bell, 11

I AM STEM, 122–23
ideas42, 156
ignorance, value of, 195–96
impact of research. *See also* actionable and policy-relevant science; policy-making, evidence-based
    about, 6–7
    Bolivian Amazon, 51–54, 57–61
    conservation biology and calls for impact, 54–57
    contact theories and, 66–68
    local knowledge dissemination and, 57–61
    public engagement, types of, 68–69
    question-asking and, 81–82
    researching research, 53–54
    Takana and Mosetén study and access points for, 61–66
incentives, 155–56, 223
An Inconvenient Truth (film), 25–27, 44
India, 33–36, 45, 46–48, 171–72
Inglis, Bob, 30
insecticides, neonicotinoid, 77–81
intentionality, 176, 203
interest groups, 214–17
Isala project, 129

Jefferson, Tom, 182, 186

Kahan, Dan, 190
Kalmus, Peter, 114–15

Kearns, Faith, 113, 178
Kimmerer, Robin Wall, 83–84
King, Natalie S., 122–23
knowledge, types of, 90–91
knowledge brokers, 218–19
Kuhn, Thomas, 85
Kwiatkowski, Richard, 207–8

laboratories, mobile, 170–71
Lacks, Henrietta, 108
Land, Michelle, 204–7, 210–11, 213–14, 220
Le Guin, Ursula K., 71
Lenfest Ocean Program, 219
LGBTQ+ rights, 47
Liboiron, Max, 106–7, 115
listening. *See also* conversations with the public
    communication and, 166
    deep canvassing and, 174–75
    "Please Just Listen to Me" model and, 14–15, 31, 34, 50
    polio vaccination listening campaigns, 34–35
    two-way dialogue and, 177–78, 222
    vaccine hesitancy and, 166
Litfin, Karen, 213
local press and media, 35, 48
local research assistants, 102–4, 117–20, 136–37
location of research, 98–101
Lugar, Richard G., 30

MacArthur, Sandy Jo, 216
Madidi region, Bolivia, 52–54, 57–58
Mainzer, Amy, 14
March for Science, 201–3
masking for COVID, 181–83, 185–90, 198–99
McIntyre, Lee, 172–73
McLaughlin, Dave, 73–75, 80–81, 90

McRaney, David, 23, 166–67
medical research. *See also* COVID-19
   Alzheimer's, 87
   Children's Heart Surgery Outcomes website, 139–43, 154
   Community-Engaged Scholarship for Health website, 155
   equity and, 101
   fetal tissue research and consent, 108–9
   Havasupai genetic samples, 102
   HIV/AIDS, 112–13
   Isala project (female microbiome), 129
   microfluidics, 110
   particulate pollution and WE ACT, 111–12
methodology, 109–13
Michaels, David, 194
Minson, Julia, 177–78
Miserendino, Rebecca Adler, 209–11, 214
miserly brain, 20–21, 43
Misztal, Barbara, 40
mobile laboratories, 170–71
Montreal Protocol, 212–13
Mosetén people, 61–66

Nagy, Chris, 124–25
Neighbours United, 174–75, 177
neonicotinoid insecticides, 77–81
Nicaragua, 117–20
No Child Left Behind Act, 148–50
norms, social and cultural, 21–22, 42–47, 102, 114

O'Brien, Mary, 93
Ojapali, 172
old-growth forests, 214–16
Oreskes, Naomi, 185, 198
over-research, 99–101

Pagel, Christina, 140–41, 154–55
Palta, Jiwan, 134–36

Palta, Monica, 8, 133–34, 136
parachute science, 95–97, 113–14
Partial Risk Adjustment in Surgery (PRAiS), 140–41
participation, public. *See also* conversations with the public
   citizen science, 128–33
   New York City Wildlife Earthwatch Research projects, 124–26
   impactful research and, 68–69
   thought-action relationship, 126–28
   unplanned, 133–37
particulate pollution, 111–12
peer review, 106–7, 145–47
Pérez Gutiérrez, Donal, 117–20, 136–37
permissions, 104–9
"Please Just Listen to Me" model, 14–15, 31, 34, 50, 223
policy-making, evidence-based. *See also* actionable and policy-relevant science
   boundary organizations and knowledge brokers, 218–20
   coalitions, 206
   engaging vs. not engaging, 217–18
   evidence-based policy movement, 148–51
   framing, 213
   interest groups and indirect approaches, 214–17
   lobbying and the policy process, 204–7
   March for Science, 201–3
   policy makers, 207–12
   politics vs. policy-making, 202–3
   windows of opportunity, 212–14
polio vaccine campaign, India, 33–36, 45, 46–48
political polarization, 30–31
Prakash, Manu, 110–11
Pratt, Mary Louise, 96
presumptive language, 42–43
"proof," 183–86
proximity to science, 170

public. *See* conversations with the public; participation, public

quadrat surveys, 184
question-asking
   about questions, 92–93
   choice of questions, 81–85
   communication, collaboration, and groupthink, 85–87
   context and types of knowledge, 88–91
   funding approaches and, 91–92
   methods, materials, and protocols, 109–13
   with farmers, 73–81, 135

Raffles, Hugh, 103
Rahman, Zia Haider, 159
randomized control trials (RCTs), 185–86
redundancy, 46–47
research, pure vs. applied, 84
research as "neutral," 96
"researching research," 53–54
research questions. *See* question-asking
right to science, 171
risk
   advocacy and, 218
   assumption of, 113
   farmers and, 74
   listening and, 159
   Partial Risk Adjustment in Surgery (PRAiS), 140–41
   permissions and consent, and, 105–6
Risk Society (Beck), 105–6
Roddick, Anita, 11
Rodman, Joey, 171
Royal Society, 89
Ruszczyk, Hanna, 100

Saab, Victoria, 215
Sagan, Carl, 195, 202
Sarewitz, Daniel, 84–85

science as verb, 120–24
science capital, 121–22
science deserts, 169
scientists, perceptions of, 49–50, 53–54, 57–61, 167–68
self-perception theory, 127–28
Sense about Science, 141–43, 209
Shah, Anuj, 216
shaming, 40–41
Shaw, George Bernard, 159
simple contagion scenarios, 43–44
skepticism and denialism. *See also* uncertainty; vaccine hesitancy
   climate change, 30–31, 172–73, 191
   COVID-19, 38, 165–66
   overconfidence and, 192
   social acceptance, 40–41, 164–166
   uncertainty, doubt, and, 194–95
Sloman, Steven, 20
Smith, Linda Tuhiwai, 53–54
social confirmation, 45–47, 164–67
social identities and information, 21–22
Solomon, Roma, 34, 223
Soulé, Michael, 54
South Atlantic Fishery Management Council (SAFMC), 129–30
specialization of knowledge, 20
Spiegelhalter, David, 193–94
Star Trek: Lower Decks (TV), 221–22, 224
Star Trek: The Next Generation (TV), 1–2, 4–5
stereotypes of scientists, 167–68
Sustainable Agricultural Research and Education (SARE), 91–92
Swift, Earl, 25, 26

Takana Indigenous nation, 60, 61–66
Tangier Island, Chesapeake Bay, 15–17, 24–29
Tantia, Piyush, 156–57
technique rebuttal, 173

Technology Policy Fellowships (AAAS), 219–20
Thacker, Naveen, 45
tipping points, social, 46–47
Tooker, John, 75–81, 91–93, 134
topic rebuttal, 173
Transforming Evidence Funders Network, 155
transparency, 193, 197–98
Trump, Donald, 17, 201–2
trust, 39–40, 48–50
Tseng, Vivian, 149–51
Tsimané-Mosetén Indigenous council, 61

uncertainty
  COVID pandemic, masking, and, 179–83, 185–90, 198–99
  data literacy and interpretation and, 187–90
  denialism fueled by certainty, 191–95
  "proof" and science, 183–86
  teaching, 195–97
  transparency, honesty, and, 193, 197–98

vaccine hesitancy
  about, 36–40
  access points and, 48–49
  COVID-19 vaccination, 36, 38–39, 192–93
  factual literature and, 41
  polio vaccine campaign in India, 33–36, 45, 46–48
  redundancy and, 46–47
  religion and, 35, 38
  shaming and, 40–41
  simple vs. complex "contagion" scenarios and, 43–45
  social conformity and confirmation, 42–43, 45–47, 164–66
  trust and, 39–40, 48–50
validity, external and internal, 147–48
values, 197–99
viral spread of behavior, simple vs. complex, 43–47
volunteers. *See* participation, public

Wason selection task, 86
Weckel, Mark, 124–25
West Harlem Environmental ACTion, Inc. (WE ACT), 111–12
West Oakland Environmental Indicators Project, 133
wildlife killing contests, 204–7, 213–14
WildMetro, 161–62
Wilson, E. O., 202
Witte, Marlys H., 195

zones of agreement, 215
Zooniverse, 128–29

About the Author

Photo credit: Scott Markle

Anne Toomey is an associate professor of environmental studies and science at Pace University, a visiting scientist at the Center for Biodiversity and Conservation at the American Museum of Natural History, and cofounder of Participatory Science Solutions LLC, a social-impact research consulting company. She lives in Sleepy Hollow, New York.